HANDS-ON SCIENCE!
112 Easy-to-Use, High-Interest Activities for Grades 4–8

Dorothea Allen

THE CENTER FOR APPLIED RESEARCH IN EDUCATION
West Nyack, New York 10995

10 9 8 7 6 5

Library of Congress Cataloging-in-Publication Data

Allen, Dorothea.
    Hands-on science! : 112 easy-to-use, high-interest activities for
grades 4–8 / Dorothea Allen.
      p.   cm.
    Includes bibliographical references (p.).
    ISBN 0-87628-906-5
    1. Science—Study and teaching (Elementary)—United States.
2. Education, Elementary—United States—Activity programs.
I. Title.
LB1585.3.A44   1991
372.3'5044—dc20
                            91-9023
                            CIP

ISBN 0-87628-906-5

**THE CENTER FOR APPLIED
RESEARCH IN EDUCATION
BUSINESS & PROFESSIONAL DIVISION**
A division of Simon & Schuster
West Nyack, New York 10995

Printed in the United States of America

**To Aleene, Christin, Kelli, Kathy, and Donny
With special thanks to Carolyn for all her help**

About the Author

Dorothea Allen is a veteran teacher and researcher who teaches Biology at Boonton High School in Boonton, New Jersey, where she also serves as Science Department Chairperson and Science Coordinator. Recently honored as Teacher of the Year in her district, she is also a recipient of the Outstanding Biology Teacher Award for New Jersey, a two-time recipient of the Outstanding Science Teacher Award of the Sigma XI Scientific Research Society of North America, and the recipient of the prestigious Presidential Award for Excellence in Science Teaching. In addition to conducting numerous science workshops for teachers, she is the author of *Research Projects in High School Biology*, *The Biology Teacher's Desk Book*, *Elementary Science Activities for Every Month of the School Year*, and *Science Demonstrations for the Elementary Classroom*, as well as numerous articles which have appeared in leading educational journals.

About This Resource

Students are eager to learn. To realize this you have only to spend a short time trying to keep pace with the youngster whose interest in the world prompts an unending stream of questions. He asks: "Why is the sky blue? Why is water wet? Why can we see through glass? Why does a ball bounce?" Children have an innate curiosity about their surroundings and the many ways in which science and technology affect their lives. They are untiring in their examination of a wiggly worm, a dead insect, a spinning top, a snowflake, or a newly hatched chick. They may spend hours in deep concentration, taking things apart, putting them back together, and trying to discover how they are made and how they work. While probing for answers to some of their questions, they uncover facets of the same or related areas, prompting new questions.

Actually, up to about the age of 12, children are observers, collectors, explorers, sorters, classifiers, and questioners. They are born scientists whose natural inquisitiveness, according to behavioral child psychologists, reaches a peak between the third and seventh grade. To keep this curiosity alive and functioning is our major concern as teachers of science.

Hands-On Science! provides you with a wide array of easy-to-use, high-interest activities that capture students' attention, pique their curiosity, and challenge them to find solutions, test theories, and make decisions. You will find activities in each chapter specifically designed to focus on one of the following important goals of science education:

- creating a science-in-action atmosphere in your classroom
- involving all students
- relating science to everyday life
- teaching science process skills
- developing science and technology literacy
- relating science to social issues
- using an integrated approach to learning
- making science personally useful
- providing individual and group activities
- preparing students for scientific and technological change
- helping students develop responsible attitudes about science.

For your convenience, each activity is organized in an easy-to-use format that includes MOTIVATION, RECOMMENDED GRADE LEVEL, STRATEGIES

INVOLVED, MATERIALS REQUIRED, and PROCEDURE. Illustrations and supplementary materials are also provided. These guidelines and materials can be used as presented or adjusted appropriately for use in a particular classroom situation. While most activities specify a hands-on approach, a few, because of the nature of the activity, feature student involvement in a teacher presentation.

Hands-On Science! encourages "sciencing" in the classroom: the emphasis is on active, hands-on learning by students through manipulation of materials, with guidelines for learning rather than direct instructions supplied by the teacher. When sciencing, students look to you to offer encouragement and to verify their observations and ideas about an investigation; you serve as a facilitator of active student learning, resisting the temptation to supply ready-made answers or explanations.

Only by turning students on to science can we achieve our goal: the involvement of all students in an excellent science program—one that will prepare not only our future scientists, but also workers who are competent in the use of new technologies, and citizens who are capable of making informed decisions regarding situations associated with scientific and technological progress.

Dorothea Allen

Table of Contents

1

Creating an Atmosphere of Science-in-Action in the Classroom

Students learn from many sources. One of the primary values of science museums as learning centers stems from the fact that the displays and working models hold great excitement and wonder for students of all ages, and that they are changed frequently to keep interest alive. The value of studying science amid exhibits of natural artifacts, living specimens, and working models—all against a background of illustrative graphs—is self-evident in the enthusiasm with which students participate in the learning activities in a science museum.

An effective learning environment in the classroom incorporates some elements of the science museum atmosphere. Amid surroundings permeated by accessibly placed science objects, displays, specimens, and manipulative materials, students are encouraged to continue giving attention to some established interests carried over from the primary grades, while initiating some new and exciting interests.

There are opportunities to include both natural and physical science "enticers" for involving students in learning science via firsthand experiences, which are followed by group interaction and/or the use of library resources.

A classroom atmosphere that sparks student interest and invites active involvement throughout the school year includes both short-term and long-range approaches:

1. Activities for the short term rely heavily on their timeliness and an eye-catching appearance that attracts attention and piques student curiosity. These activities are most effective when they focus on an area of current

1

study or on a scientific event that is seasonal or newsworthy, and when they are changed frequently.

- *Posters and bulletin board displays* keep students alerted to science events and to "Science in the News."
- *Mobiles* depict the changing positions of objects in a system, such as the planets in a miniature solar system or the relative positions of the earth, sun, and moon at specific times of the day, month, or year.
- *Scale models* are useful for enabling students to perceive the proportions of component parts, such as the comparative length of arms and legs possessed by primitive man or the wingspread of an eagle.
- *Dioramas* are used to portray a scene in a specific locale, such as a prehistoric setting featuring dinosaurs and giant ferns or a space station with possible forms of life.
- *A curiosity corner* provides a place for students to examine unusual rocks, shells, leaves, or other specimens that have been brought in by others for sharing with classmates, or to become familiar with an unusual device such as a robotic toy.
- *Working models* of a human heart, an arm, or a leg, provide opportunities for students to examine the working parts and to observe some structure/function relationships.

2. Activities for the longer range are versatile if maintained in basic form and adjusted appropriately to accommodate new points of focus and/or student involvement over a period of time. This applies directly to "living science" centers in the classroom that attract attention and encourage student participation in some exciting and challenging learning experiences.

CLASSROOM ACTIVITIES

1–1 MAINTAINING A FRESHWATER AQUARIUM IN THE CLASSROOM

Motivation: An aquarium is a primary focal point of interest in the science-oriented classroom. It is an immediate attention-getter that both generates and sustains interest in the inhabitants and their activities as they play out the drama of life in an aquatic community. Long-range considerations favor a simple design that can be modified to accommodate new points of emphasis in an ongoing study of the living environment, with opportunities also for students to make suggestions and contributions from time to time.

Recommended Grade Level: Grades 4–6

Strategies Involved: Student involvement
Science skills development

Materials Required:

- a low, rectangular aquarium tank with a capacity of 10 or more gallons
- washed sand
- clear pondwater or tap water that has been aged for 24 hours
- aquatic plants of one or more varieties, such as *Elodea, Cabomba, Sagittaria,* or *Vallisneria*
- several freshwater fish of compatible types
- metal strips
- plastic bags

Procedure:

Enlist the aid of student volunteers for performing the following:

1. Place an aquarium tank, or other container that provides a large surface area for contact with the atmosphere, in an area that provides sunlight and allows for clear viewing by students.
2. On the bottom of the tank, place washed sand to prepare a base layer that tapers from 2.5 cm at the front to 5 cm at the back.
3. Fill the tank with clear water, taking care not to disturb the sand.
4. Attach metal strips to the lower ends of selected aquatic plants and anchor these plants in well-spaced positions in the sand.
5. Allow the sand to settle until the water is clear and its temperature is about that of the classroom.
6. Place a small number of freshwater fish, selected for transfer to the aquarium, in a plastic bag containing water from the original habitat.
7. Float the plastic bag and its contents on the surface of the water in the aquarium tank.
8. After one or two hours, when the temperature of the water in the plastic bag is the same as that of the aquarium water, release the fish into the water in the tank.

9. Establish a regimen for fish-feeding, so that students can observe the feeding responses made by individual specimens, as well as by the population as a whole.

Encourage students to view the aquarium as a model of a natural habitat, with attention directed to interrelationships between its various members and between the inhabitants and factors of their physical environment, focusing on:

1. the need for a tank that provides a large surface area for the water,
2. the purpose of plants in a "fish" aquarium,
3. the effects produced by adding more fish of the same or different species, or of other specimens such as snails, small crayfish, or "algae-eaters."

1-2 NURTURING RESIDENTS OF A BUTTERFLY GARDEN

Motivation: The emergence of a butterfly from its chrysalis is an exciting climax to a classroom activity that follows the developmental stages in the life of an insect. When placed in a container that allows for easy viewing by all students, butterfly larvae collected from leaves on native plants can be observed over a period of months, during which time they feed, grow, prepare their pupa cases, and withdraw from visible activity until the long-awaited E-Day. Students should be encouraged to "adopt" a larva and follow the sequence of events in which the organism undergoes a complete change in form and emerges as an adult butterfly.

Recommended Grade Level: Grades 4–6

Strategies Involved: Student involvement
Science skills development

Materials Required:

- a clear plastic sweater box or shoe box
- a shallow cup
- a square of nylon net
- moist soil
- a large rubber band
- two or three caterpillars of painted lady or other species
- two or three leafy twigs from the plant on which the caterpillars were found

Procedure:

Enlist the aid of student volunteers for performing the following:

1. Place a 1-inch base layer of moist soil in the bottom of a clear, plastic sweater box or shoe box.
2. Set a small cup of water on the surface of the soil.
3. Place two or three leafy twigs so that their cut ends are immersed in the water in the cup and their upper ends rest against the side of the box.
4. Gently place the selected caterpillars on the twigs.
5. Place a square of nylon net over the top of the box and secure it with a strong rubber band, thus forming an escape-proof lid that also provides the caterpillars with proper ventilation.

Encourage students to develop a sensitivity to the events occurring in the butterfly garden and to the well-being of the inhabitants. They should:

1. Make necessary adjustments to ensure proper conditions of air, light, temperature, and moisture for the developing organisms.
2. Observe the gradual changes that occur as the caterpillars pass through successive stages of development in their life cycles, and mark on a calendar the amount of time spent in each developmental stage.
3. Release adult butterflies into the outside environment when weather conditions are suitable.

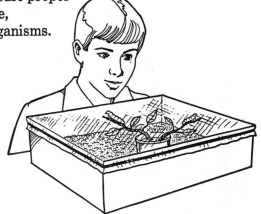

1-3 OBSERVING ACTIVITY IN AN EARTHWORM FARM

Motivation: Earthworms do more than crawl over the surface of the ground at night. Most of their time is spent below ground level, where they avoid light and dig tunnels by pushing excess soil out of the way or by passing it through their bodies and depositing it in the form of worm castings near the entrance to their burrows. Sometimes they do come out of their burrows during the day when the air spaces in the soil become filled with water. Students can observe these below-ground activities when earthworms are placed in an earthworm farm whose transparent walls have portable covers for maintaining conditions of darkness during all but periods of actual viewing.

Recommended Grade Level: Grades 4–6

Strategies Involved: Student involvement
Science skills development

Materials Required:

- a large glass jar
- earthworms from a flourishing vegetable or flower garden

- rich, moist soil from the garden where the earthworms were collected
- an empty soup can, smaller than the jar and closed at one end
- black construction paper
- tape
- sand
- water
- an atomizer or "mister"
- grass cuttings and/or leaf fragments

Procedure:

Enlist the aid of student volunteers for performing the following:

1. Remove one end from a clean soup can.
2. Place the can, closed end up, in a large glass jar.

3. Place rich, moist garden soil in the glass jar, completely filling in the area around the can and covering the top of the can with a thin layer so that the can is no longer visible.
4. Add a thin layer of sand to the top of the soil in the jar.
5. Scatter leaf fragments or grass cuttings over the surface.
6. Moisten the surface lightly, using an atomizer or a "mister."
7. Transfer earthworms to the jar, setting them gently atop the moistened surface.
8. Cover the outside of the jar with black construction paper, secured to the jar with tape to make it light-tight.
9. Set up a schedule of regular "mistings," to be performed daily by involved students.

After light-tight conditions have been maintained for a period of 24 hours, allow students to remove the dark paper cover for a brief period of viewing the earthworm farm. Observations and comments should focus on:

- evidence of tunneling,
- evidence of worm castings,
- reasons for the black paper cover,
- necessity for maintaining proper conditions of moisture,
- contributions of earthworms to a natural garden setting,
- reasons for replacing the black paper cover after periodic viewings.

1–4 PLANTING A WOODLAND TERRARIUM

Motivation: Placing a woodland terrarium on display in the classroom has some far-reaching effects: students develop a greater awareness of the organization within a natural setting and are provided with an incentive to prepare other models representing settings in desert, bog, or tropical environments. Prepared as small group or whole class projects, or by individual students, these simulations of real environments provide for a comparative study of the different physical settings and their ability to provide the requirements for life of the inhabitants. When properly planned and maintained, terraria will last for many months, during which time students can observe many natural cycles as well as some interrelationships between the organisms that live in a common environment.

Recommended Grade Level: Grades 4–6

Strategies Involved: Student involvement
Science skills development

Materials Required:

- a clear glass or plastic container with a cover of wire netting
- plant specimens, such as moss, fern, and small seedlings of native woodland trees and/or shrubs
- small animal specimens, such as newts or salamanders
- potting soil
- charcoal pieces
- pebbles, gravel, or shell chips
- small rocks
- a small trowel
- a watering spray bottle
- a metric ruler
- a notebook or some other provision for record-keeping

Procedure:

Enlist the aid of student volunteers for performing the following:

1. Select an attractive glass container that is of suitable size and shape for the specimens to be included.
2. Prepare a 2.5- to 5-cm layer of pebbles or other drainage material in the bottom of the container.
3. Spread a layer of charcoal pieces over the pebble layer.
4. Cover the charcoal layer with a layer of potting soil, building the terrain higher in some areas than others.
5. For interest and variety, position one or more rock specimens in strategic locations.
6. Moisten the soil and place plants appropriately to create a natural looking setting: small sprigs of club moss and mountain laurel in front, taller ferns

and seedlings at the back, woodland moss and partridge berry as an allover ground cover, and scarlet *Cladonia,* variously placed for added color.

7. When all plants are in place, introduce a few newts and/or salamanders to complete the woodland terrarium.

8. Place the cover on the terrarium and position the entire assembly in a location that provides good quality light and allows for easy viewing by students.

Involve students in all care and maintenance procedures:

• Remove the cover daily to check the moisture conditions and apply appropriate adjustment measures, as needed—spray the plantings with a fine mist or allow the condensation to escape from the terrarium before replacing the cover.

• Remove all dead and decaying organic matter.

• Replace living specimens that are no longer in a healthy condition.

• Trim excess plant growth to prevent direct contact of plant tissue with the top and/or sides of the planter.

Engage students in an open class discussion that focuses on their discovery of some important natural phenomena being played out in the miniature model:

• Growth activities of plants are represented by an increase in size of the seedlings and by the appearance and uncurling of young fern frond "fiddle-heads."

• Survival abilities of animals are illustrated by the salamanders' selection of moist hiding places under the rocks.

• The hydrologic cycle in nature is repeated in miniature when evaporation and condensation of moisture is detected within the terrarium setup.

Encourage students to investigate the availability of materials for other types of terraria and to make plans for the preparation of bog, desert, and/or tropical environments that will serve as a comparative study.

2

Involving
All Students

Middle-grade students are not merely older than they were as primary school youngsters; they are now students whose interests have become more selective, whose attitudes now extend beyond a self-centered focus, and whose capacity for developing an awareness of the varied investigative methods and their uses is continually growing. At this critical level, care should be taken to avoid the development of student boredom or frustration by the treatment of this period as an interim, either for repeating with greater precision some previously performed activities or for preparing for secondary school activities. Clearly, the middle grades are more than a transition between the lower elementary grades and the high school, and students in these grades are worthy of a science program that addresses their specific interests and needs.

A science program for the middle grades needs to focus on the students as unique individuals—to help them broaden their views, to cultivate their interests, to encourage their enthusiasm for learning, and to increase their feelings of achievement as they seek answers to questions about the world around them. Maintaining student interest and involvement at a high level is necessary to achieve these goals. An effective program involves the use of approaches that encourage student input in maintaining a science atmosphere in the classroom, provide incentives for students to expand their interest and involvement in science, and involve all students as active members of a group participating in a variety of learning experiences.

ACTIVITIES THAT MAINTAIN AN ATMOSPHERE
OF SCIENCE-IN-ACTION

2–1 PROVIDING WATER FOR USE BY NEWLY HATCHED CHICKS

Motivation: Students show great concern for the well-being of animals being maintained in the classroom, and give top priority to providing them with food and water. They soon realize, however, that newly hatched chicks that have been placed in a brooder are unable to use the kind of water bottle that is used so successfully by hamsters and guinea pigs in an animal cage. Nor is a small saucer a proper vessel to use, because the chicks can be observed to overturn the saucer and spill the water or to run through and foul what water remains in the saucer. Assembling a drinking trough that makes available a steady supply of water and does not pose a threat of death by accidental drowning holds a built-in motivation for students to incorporate some basic scientific knowledge into a practical design that will solve the problem.

Recommended Grade Level: Grades 5–8

Strategies Involved: Hands-on activity
Student involvement
Science skills development

Materials Required:

- a shallow, circular pan
- modeling clay
- a mayonnaise jar with a screw-cap cover

Procedure:

Enlist the aid of student volunteers for performing the following:

1. On the inside bottom of a shallow pan, trace the outline of the outside edge of the open end of the mayonnaise jar. Then position three lumps of modeling clay at points that are equidistant from each other along this circumference.
2. Next, pour water into the pan until it is three-fourths full of water.
3. Completely fill the jar with water and place its cover in position. Then, invert the jar in the water-filled pan and, while holding it underwater, remove the jar cover, allowing no water to escape from the jar.

4. Keeping the mouth of the jar below the surface of the water in the pan, ease the jar to a position where its rim rests on the clay supports. Then transfer the intact water trough to the chick brooder.
5. Set up a schedule for students to prepare a fresh water trough on a daily basis.

Provide opportunities for students to observe the chicks drinking from the pan and the mechanism that provides a constant supply of water, noting:

- a gradual increase in the height of the air column that forms in the jar above the water,
- a gradual lowering of the water level in the jar.

Conduct an open class discussion in which students consider key points:

- the advantage offered by the use of a water trough of this design, rather than simply a pan of water,
- the purpose served by the clay supports in permitting a transfer of water from the jar to the pan,
- the identification of this mechanism as a modified siphon system, which remains operable as long as the water level in the jar is higher than the water level in the pan.

2-2 MAKING USE OF A SIPHON SYSTEM

Motivation: Classroom aquaria are most successful when they simulate conditions of the natural environment as much as possible. In addition to factors of food, light, temperature, and oxygen supply, the condition of the aquarium water is of primary importance. Student involvement in a schedule for changing the aquarium water helps to guard against pollution buildup, while employing a scientific technique in a new and practical situation.

Recommended Grade Level: Grades 4–8

Strategies Involved: Hands-on activity
Student involvement
Science skills development

Materials Required:

- an active aquarium
- clear plastic tubing
- a collecting bucket
- a supply of fresh pondwater
- a holding tank or bucket
- plastic sheeting or paper toweling
- a low stool

Procedure:

Enlist the aid of student volunteers for performing the following:

1. Transfer all living forms from an aquarium that needs to be cleaned to a holding tank for temporary housing.
2. Place plastic sheeting on the floor below the table on which the aquarium is located. Set a low stool on the plastic sheeting beside the table and place a collecting bucket on the stool.

3. Immerse a length of plastic tubing in the aquarium water and allow the tube to become completely filled with water. Make certain that no air remains in the tube. Then, using one finger (or the thumb) of each hand, cover both ends of the tube to prevent the escape of any water.

4. Keeping one end of the tube underwater in the aquarium tank, hold the other end of the tube above the collecting bucket that is in place beside the table. When you are ready to allow the water to begin to flow, remove your fingers from the ends of the tube.

5. Holding the tube in position, make adjustments, as necessary, to keep one end submerged in the aquarium water and the other directly above the bucket on the stool, until the desired amount of water has been transferred.

6. Replace the water removed from the tank with fresh pondwater.

Encourage students to observe closely and trace the flow of aquarium water as the siphon system operates:

- upward in the tube from the aquarium tank,
- downward in the tube to the collecting bucket on the stool.

Conduct an open class discussion in which students focus attention on specific aspects of the design and operation of the siphon system:

- the need to replace air in the tube with water when setting up the siphon system,
- the constant streaming of water through the tube as long as the water level in the aquarium tank is maintained at a higher level than the lower end of the tube, which is directed toward the collecting bucket, and as long as no air is allowed to enter the tube.

As a follow-up, have each student prepare a record of the activity, including:

- a diagram that shows the pathway taken by water being transferred from the aquarium to the collecting bucket,
- a description that indicates how a knowledge of gravity and air pressure is helpful when setting up and operating a siphon system.

2-3 MAINTAINING A SUITABLE LEVEL OF HUMIDITY IN CLASSROOM TERRARIA

Motivation: The atmosphere in a commercial greenhouse is both hot and humid. These conditions, which promote the highly desirable lush growth of foliage, are maintained on a year-round basis, despite the weather conditions out-of-doors. To protect the plants from the drying effects of high heat, watering techniques that include the spraying of a fine mist are employed. In a similar manner students can maintain proper conditions of humidity by "misting" plants in a classroom terrarium, using a spray bottle whose design and operation are based on familiar scientific principles.

Recommended Grade Level: Grades 4–6

Strategies Involved: Hands-on activity
                           Student involvement
                           Science skills development

Materials Required:

- a spray bottle from a household cleaning product
- rainwater
- a woodland terrarium in the classroom

Procedure:

Enlist the aid of student volunteers for performing the following:

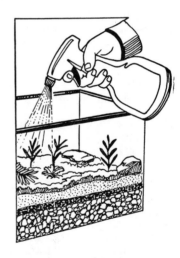

1. Thoroughly clean an empty spray bottle from a window cleaner or other household product.
2. Fill the bottle with rainwater.
3. Direct the delivery end of the spray mechanism toward plants in the terrarium and, using a pumping action, operate the spray bottle.

Provide an opportunity for students to examine the spray pump closely and to analyze the mechanism in operation. Attention should focus on:

- the size of the outlet at the top of the bottle as compared with the diameter of the tube extending down into the water,
- the pumping action, which produces a partial vacuum in the tube and compresses the air in the space above the water level in the bottle,
- the increase in pressure on the surface of the water in the bottle, resulting in the water being forced upward in the tube,
- the release of water through a narrow outlet at the top of the bottle, causing the water to be broken up into small particles and resulting in a "spray."

- the increasing difficulty experienced in the operation of the system as the water level in the bottle is lowered and the amount of air to be compressed in the space above the water is increased.

Allow students to meet in small discussion groups for considering the benefits enjoyed by the terrarium plants being "misted." Then, have each group prepare:

- a list of ways that this method of creating a moist environment for houseplants patterns itself closely after nature's way,
- a summary statement that compares the "misting" of houseplants with the more common practice of watering that relies solely on applying water to the soil.

2–4 BALANCING PAPER CUTOUTS FOR A MOBILE

Motivation: Suspending an object at a point that is midway between its two ends does not always produce a balanced system. For example, when preparing a mobile using irregularly shaped cutouts, such as those representing a collection of bird, fish, reptile, or mammal figures, it will be necessary to find the balance point for each figure, just as two students must find the point at which each must sit on a seesaw in order to effect a balanced system. Finding this point by a trial-and-error method often proves to be a time-consuming and frustrating experience. However, finding it by a method that involves students in an intriguing aspect of science introduces them to a useful technique for finding the balance point of any irregularly shaped cutout figure.

Recommended Grade Level: Grades 5–8

Strategies Involved: Hands-on activity
Student involvement
Science skills development

Materials Required:

- paper cutouts of a collection of birds, butterflies, fish, or other irregularly shaped cutout figures
- strong thread
- chalk
- a pencil or other marker
- a ruler
- a bulletin board or other supporting structure
- large, sturdy straight pins

Procedure:

Instruct students to proceed as follows:

1. At a convenient height, stick a straight pin into a bulletin board, embedding it deeply, but allowing it to protrude about ½ inch at the head end.

2. On the back of an irregularly shaped cutout figure, mark two positions that are at a fair distance from each other and are located close to the outside edge of the figure.

3. Using a straight pin or a small nail, punch a hole through the cutout at both locations marked and label them #1 and #2, respectively.

4. Construct a plumb line and bob by attaching a large paper clip to one end of a strong thread and fashion a loop at the opposite end for slipping over the projecting pin shaft on the bulletin board.

5. With the front of the paper cutout against the bulletin board, suspend the cutout by easing hole #1 over the pin shaft which protrudes from the bulletin board.

6. Place the thread loop over the pin so that the thread will hang freely, weighted by the paper clip at its lower end.

7. On the back of the cutout, place a chalk mark to indicate a point where the thread rests against the paper cutout near its lower edge.

8. Remove the assembled plumb line and bob and the cutout from the bulletin board. Then, use a ruler to draw a line segment that connects hole #1 with the chalk mark.

9. Repeat steps 5–8 for hole #2. Then mark the point at which the two line segments cross. This point is called the *balance point*.

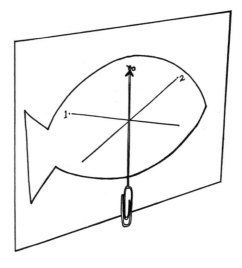

10. Determine the position you desire the figure to take on the mobile. Slip the loop of the plumb line and bob over the shaft of a pin and, holding the figure against the bulletin board in the desired position, move the plumb line along the figure until the plumb line crosses the balance point. Stick the pin through the figure to hold it in this position on the bulletin board.

11. Remove the figure from the bulletin board and draw a line segment from the balance point to the hole made by the pin. Now punch a hole along this line segment on the cutout near its upper edge. Attach a length of thread to this hole. The figure is now ready to be suspended in its proper position on a mobile.

As a follow-up, engage all students in an open class discussion that focuses on:

- an estimate of the number of trials that might be necessary for finding the balance point of a cutout figure, using a trial-and-error method,
- the relative efficiency and accuracy of scientific and trial-and-error methods in scientific work and in everyday situations,

- reasons why the balance point of an object is also referred to as its *center of gravity.*

ACTIVITIES THAT EXTEND STUDENT INTEREST

2–5   OBSERVING THE DISTRIBUTION OF OBJECTS ACCORDING TO SIZE

Motivation: Finding only large blueberries at the top of a berry basket can give some false impressions—usually the basket contains small berries as well, but they are found at the bottom. One experiences a similar situation with a bag of potato chips; only small bits and pieces of broken chips remain in the bottom of the bag after the large, whole chips have been removed from the top and eaten first. In both instances, the findings seem to be contrary to expectations that gravity, having a greater attraction for larger, heavier objects, should, in each case, cause the larger berries/chips to be situated below the smaller, lighter ones. Involving students in an activity that investigates this seemingly contradictory situation can be accomplished with some simple materials in the classroom.

Recommended Grade Level: Grades 5–8

Strategies Involved: Student involvement
Science skills development

Materials Required:

- a transparent plastic box
- a supply of dried pea seeds or colored beads
- a tennis ball or other ball of a contrasting color

Procedure:

Involve the entire class in the activity, as follows:

1. Allow students to examine the tennis ball and the dried pea seeds or colored beads to be used in the activity.
2. After students have viewed and handled the objects, ask them to predict which will be located at a lower level if both are placed in the same container.
3. Place the ball in the center of the plastic box and pour beads into the container, covering the ball completely.
4. Then, shake the container from side to side while students observe what happens to the objects in the box.

Encourage students to check the correctness of their predictions and to analyze the demonstrated events in which the ball is nudged to the top, where it comes to rest atop the beads:

- The shaking action allows the small size beads to drop down into the spaces at a lower level and to nudge the larger ball to the top.
- The rearrangement of objects in the container causes a lowering of the center of gravity, which results in the system becoming more stable.

Reinforce student learning by asking students to relate the activity to some personal experience that comes to mind, such as:

- finding the largest blueberries on the top layer of a berry basket, with smaller ones at the bottom,
- finding cookie crumbs at the bottom of a bag or box of cookies in which some of the cookies have been broken.

2–6 BLOWING BUBBLES IN WINTER

Motivation: Blowing bubbles has become more than a childhood pastime. It is now a popular attraction at some of the major science museums and science activities centers throughout the country. Recently it was also a real attention-getter on the plaza of the IBM building in Manhattan, where businessmen and women, distracted from their lunchtime activities, lingered to take a turn at bubble blowing and to participate in animated discussions with each other and with equally intrigued passersby.

Students engaging in a bubble-blowing activity derive great pleasure from this encounter with real science. By extending the basic activity to include different bubble solutions, different extruders, and different atmospheric conditions, students have an opportunity to raise pertinent questions and to explore the reasons for bubbles of different size, formation, and color, and for the different heights to which the bubbles are observed to rise.

Recommended Grade Level: Grades 4–8

Strategies Involved: Hands-on activity
Student involvement
Science skills development

Materials Required:

- a variety of bubble mixtures (such as ⅔ cup of dishwashing liquid in 1 gallon of warm distilled water)
- a pan of suitable size for each bubble mixture used
- an assortment of bubble pipes, drinking straws, and/or wire loops

Procedure:

On a clear, cold day in winter, assign a group of students to transfer warm bubble mixtures and an assortment of bubble pipes out-of-doors, where the entire class can engage in a bubble-blowing activity. Each student should:

1. Inhale a lungful of air and hold it for a few moments.
2. Dip a bubble pipe into a warm bubble solution and use the air being held in the lungs to blow a bubble.

Encourage students to use a variety of bubble mixtures, as well as a selection of tubes and wire loops of different shapes, for blowing additional bubbles, to be compared in relation to their

- size,
- shape,
- color,
- activity.

Ask students to relate some experience in science to the observations they make and to the questions they raise concerning:

- reasons that would account for differences noted in bubbles blown in summer and in winter,
- the shape of bubbles blown by differently shaped bubble pipes and loops,
- the occurrence of bubbles that bounce and the formation of clusters of bubbles that may have been observed.

2–7 SHARPENING POWERS OF OBSERVATION FOR SCIENCE

Motivation: It is not uncommon for several people who have witnessed the same event to report different accounts of what happened. Clearly, the skill for making careful and detailed observations is not well-developed among most people. Students need to have opportunities for developing an appreciation of the importance of observation as a useful process of science and they must have incentives for improving their individual skills, where necessary. This can be initiated by an activity that allows them to check the degree of accuracy and attention to detail of their own observations, which may be useful to them in solving a problem or, if faulty, may lead them to a wrong conclusion.

Recommended Grade Level: Grades 4–8

Strategies Involved: Student involvement
Science skills development

Materials Required:

A tray with 12 assorted objects:
(colorful seashell, rock specimen, eraser-topped pencil, mounted butterfly, funnel, pinecone, Magic Marker, chalkboard eraser, medicine dropper, piece of colored chalk, two rulers of different length)

Procedure:

Plan and conduct an activity that
involves all students:

1. Keep the prepared tray covered and
   out of sight until it is to be used.
2. Allow students to observe the tray as
   it is passed around the classroom or
   displayed on a science exhibit table.
3. Remove the tray from the area of
   student viewing.
4. Ask the students to list in their individual science notebooks all of the
   objects that were viewed on the tray.

Ask students to volunteer information about the objects they have listed and to
place the names of these objects on the chalkboard. Then, when no more contributions are offered, reveal the tray so that students can check the accuracy and
completeness of:

- their individual lists,
- the list on the chalkboard prepared by the entire class.

Engage students in a class discussion that focuses on their response to specific
questions:

- How many students listed all 12 objects correctly?
- How many students forgot to list two or more of the objects?
- How many students listed something incorrectly or included something that
  was not on the tray?
- How many different groups of objects can you identify in the 12 objects on
  the tray?
- Does it appear that it is useful to make groupings and/or associations when
  making observations?

Students may report personal experiences involving success with the technique
of grouping things by size, color, usage, or some other characteristic that might
reasonably account for their being declared "careful observers" in the activity.

They should apply the experience to situations in which reporting observations
completely and accurately contributes to success in science and to situations in
which different accounts are reported by three students who witnessed the same
disturbance in the school lunchroom.

2–8   USING SCIENCE INFORMATION TO SOLVE A MYSTERY

Motivation: Most students will recall one early childhood toy that was different
from all others—unlike most toys that had to be picked up and placed in an upright

position after having been knocked down, the unusual toy clown would always return automatically to an upright position, no matter how many times or with what force it was pushed over. Viewed from the vantage point of their now greater maturity, students in the middle grades can renew their acquaintance with this toy, replacing the frustration they experienced in early childhood with a conscious curiosity that involves them in an activity that investigates WHY the toy clown acts as it does.

Recommended Grade Level: Grades 5–8

Strategies Involved: Student involvement
                          Science skills development

Materials Required:

A familiar childhood toy clown that always returns to an upright position when pushed over

Procedure:

Conduct a class activity that involves all students:

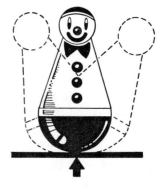

1. Place the toy clown on a table top where all students can view it clearly.
2. Push the clown backward, forward, and to either side, each time releasing the toy before toppling it in a different direction.
3. Invite students also to push the clown over in whatever direction they choose, while observing what happens when the toy is released after having been pushed over in each direction.
4. Allow students to examine the toy and its design and construction.

Encourage students to probe into the scientific principle that explains the clown's observed behavior, using as helpful guidelines for orderly thinking their answers to a series of questions:

- Does it appear that the entire toy was constructed of the same material?
- Which part of the clown's body appears to be made of a denser material?
- Does the ball-shaped base appear to be solid?
- Where does the heavier material in the base appear to be concentrated?
- What effect does this have on the balance of the toy clown?
- Where do you conclude the center of gravity of the toy clown to be located?
- How does this information help to solve the mystery of how the clown always springs back to an upright position after being toppled?
- Was the information needed to solve the mystery collected by direct or indirect observations, or by a combination of the two?

2-9   REVEALING THE INSIDE STORY OF A BIONIC BIRD

Motivation: A glass-and-plastic model of a bird that lowers its head into a glassful of water, and then raises it again, gives the appearance that the "bird" is actually drinking the water. Once started, the head-bobbing action is repeated over and over again in a motion that resembles that of a seesaw. Close examination of the model's construction and careful observation of its operation will stimulate student thinking about the science-based changes that are involved and how they contribute to the seemingly constant motion. Involvement of students in the bionic bird activity sharpens their awareness that science can be fun and that an understanding of science increases the enjoyment they can derive from many science-based novelties.

Recommended Grade Level: Grades 7–8

Strategies Involved: Student involvement
Science skills development
Science enrichment

Materials Required:

- a drinking bird demonstration model*
- a glass which is almost full of water

Procedure:

Involve all students in a class activity, as follows:

1. Place a glass which is almost full of water on a table where all students can view it clearly.
2. Place the drinking bird model next to the glass.
3. Carefully, duck the bird's head into the water in the glass, and then restore the bird to its upright position. (NOTE: The room temperature must be maintained above 65 degrees F.)
4. Allow sufficient time for students to observe the repeated action of the bird, as it alternately dips its head to "drink" and raises it between sips.
5. Instruct students to examine the construction of the model and to analyze its action by focusing on important considerations:

   - the ease with which the colored liquid[†] in the tube-like body passes from one end to the other,
   - the effect on vapor pressure in the head region, caused by the evaporation of water from the surface of the wet head each time it emerges from the water in the glass,
   - the change in vapor pressure when gases in the head region are cooled,

*Available at nominal cost from most science supply companies.
[†]Methylene chloride, a volatile liquid that evaporates readily at room temperature.

- the pressure differences that force the liquid to rise in the tube,
- the shifting of the center of gravity, which causes the bird to dip its head in the water again,
- the causes and results of changes in vapor pressure,
- the importance of maintaining the room temperature at levels above 65 degrees F.

Provide for reinforcement of students' understanding by having them prepare a summary of the activity in which they identify the science involved in the sequence of events of one complete cycle:

- In warm surroundings, some of the volatile liquid residing in the bird's lower body evaporates, creating an increase in vapor pressure on the surface of the liquid and forcing some of the liquid to rise in the tube and enter the head chamber.
- When the weight of the liquid forced into the head is great enough to raise the bird's center of gravity, the bird tips forward and dips its beak into the water in the glass.
- In this position, the bottom of the tube is above the surface of the liquid, which allows the liquid to spill back into the lower part of the bird's body. The center of gravity is thus lowered and the bird is restored to its upright position.
- Water evaporating from the wet beak cools the head, reducing the pressure in that area and allowing the vapor pressure in the bird's body to force the liquid up the tube once again.

ACTIVITIES THAT PROVIDE A VARIETY OF LEARNING EXPERIENCES

2–10   OBSERVING MOVEMENT IN A NONLIVING FORM

Motivation: The spontaneous movement exhibited when a bird flies or a worm crawls is usually associated with activities engaged in by living organisms. It is, in fact, one of the basic considerations in distinguishing between living and nonliving forms. An activity that involves students as they observe movement in a nonliving object allows them to analyze the situation and to suggest possible reasons for the type of movement observed here and in some other everyday encounters.

Recommended Grade Level: Grades 4–6

Strategies Involved: Hands-on activity
Student involvement
Science skills development

Materials Required:

Each group of four students will need:

- soda straws with paper sleeves
- a medicine dropper

- a paper cup containing a small amount of water
- a small saucer or glass plate

Procedure:

Instruct students, working in groups of four, to proceed as follows:

1. Carefully tear off the top end of the paper sleeve covering of a soda straw.
2. Place the straw in an upright position on a table top so that the open end of the sleeve is uppermost and the straw within is just visible.
3. Slide the paper sleeve down the length of the straw, crunching the paper to a length of about 1 inch.
4. Remove the straw and place the empty, crunched paper tube on a saucer or glass plate.
5. Using a medicine dropper, scatter a few drops of water along the crunched paper tube.
6. After viewing the paper "worm," compare its movement with that of a real worm, noting:

- ways in which the movement of the paper resembles that of a worm,
- ways in which the movement of the paper is different from that of a worm.

2-11   DISCOVERING THE SECRET OF A MAGIC VIEWER

Motivation: There are many reasons for viewing an object through a substance that is transparent; the object may be magnified, the focus and detail of a very distant object may be sharpened and more clearly defined, or some interfering wavelengths of light may be filtered out to produce a more desirable effect. Sometimes these changes are accompanied by additional effects; there may be distortions or, in some cases, due to the bending of light rays as they pass from one medium to another, the image produced may appear in an inverted position. A changed appearance of only one word of a two-word phrase being viewed through the same "magnifier" will challenge students and encourage them to concentrate and analyze the situation as they work in small groups to find an explanation for this phenomenon.

Recommended Grade Level: Grades 5–8

Strategies Involved: Hands-on activity
Student involvement
Student interaction
Science skills development

Materials Required:

Each group of four students will need:

- one large glass test tube with stopper to fit
- lined file cards or paper
- two Magic Markers of contrasting color
- water

Procedure:

Instruct students, working in groups of four, to proceed as follows:

1. Prepare a "magic viewer":
   - Completely fill a test tube with water.
   - Insert the stopper in the mouth of the tube and check to be sure that it is watertight.
2. Prepare a piece of lined paper or a file card with selected words for viewing. With upper case letters printed neatly one space high and using a different color for each word, print CODE WARP or other suitable words for use with the magic viewer.
3. Holding the tube sideways, about 1 inch above the lettering on the card, bring the letters into focus and view the two words.
4. Discuss with members of the group the appearance of the two words and investigate the matter further to discover an explanation for the unusual effect:

   - Rest the tube directly on top of the lettered card and view the words again. Then, slowly raise the tube and locate the distance at which the letters become inverted.
   - Consider whether the tube can invert some letters and not others.
   - Look for other possible explanations (color, size, letters that are vowels or consonants, symmetry of the letters, or other feature) that apply to the letters of the words being viewed.

Encourage students to plan and conduct extensions of the investigation in which they:

- prepare additional cards for viewing, using upper case letters to spell out original word groups (PARTY HEX, FAULTY CHOICE, DECODE STAMP, and others), in which only one of the words is made up of letters having the same appearance, whether viewed with the naked eye or through the "magic viewer."
- design and conduct an investigation for determining if the effects produced by the "magic viewer" are a result of light passing through the glass walls of

the tube, through the water, or a combination of the two, and report their findings to the class.

2-12 DETECTING AND IDENTIFYING DUPLICATE OBJECTS THAT CANNOT BE SEEN

Motivation: Everyone has experienced a certain curiosity about the identification of something he/she cannot see. A gaily wrapped birthday gift, for example, will conjure up some ideas about what is contained in the package, with consideration given to the size, weight, and shape of the package, as well as to possible sounds that are produced from within. An activity that asks students to identify objects they cannot see encourages them to use a similar approach: using information gathered by indirect methods enables them to "see" what is inside a closed container, and to make corresponding associations with common objects with which they are familiar, in order to arrive at a reasonable identification.

Recommended Grade Level: Grades 4–6

Strategies Involved: Hands-on activity
Student involvement
Student interaction
Science skills development

Materials Required:

- a supply of cylindrical, paperboard pint containers with lids (or cleaned-out coffee containers from takeout orders)
- a supply of five different small objects, such as dried lima bean seeds, mothballs, marbles, buttons, and paper clips

Procedure:

Plan and conduct an activity in which all students, working in groups of four, become actively involved:

1. Prior to classtime, prepare six containers for each group of four students:

   - Place six dried lima bean seeds in an opaque container that can be closed securely with a lid.
   - Similarly, prepare four additional containers, using the remaining four kinds of objects.
   - Prepare one container that duplicates any one of the five containers already prepared.
   - Label the containers, randomly, with #1–#6.

2. Distribute one set of six prepared containers to each group of four students.
3. Present the mystery of the contents of the containers and allow each group to plan a procedure for collecting and recording useful information that does not permit the opening of the containers.

4. Allow time for each group to conduct an investigation, to analyze the information (weight, movement, sounds, and other characteristics) collected, and to make decisions concerning:

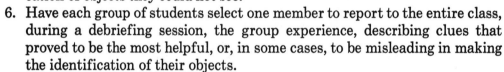

- which two containers hold the same kind of object,
- the identification of the objects in the containers holding duplicates,
- the identification of the other objects being held in other containers.

5. Permit students to open their containers and verify the accuracy of their identification of objects they could not see.

6. Have each group of students select one member to report to the entire class, during a debriefing session, the group experience, describing clues that proved to be the most helpful, or, in some cases, to be misleading in making the identification of their objects.

Then, as a follow-up activity, engage all students in a discussion of the activity, focusing on the many ways of collecting information about objects that are alike and those that are different.

2–13   CHECKING OUT MONEY THAT JUMPS

Motivation: There is much evidence of the effect that heat has on some well-known objects; a steel building is taller in summer than in winter, spaces between adjoining sections of concrete roads and sidewalks are narrower in summer than in winter, automobile and bike tires appear slightly deflated the morning after a cool night, even though the tires had been fully inflated the previous warm afternoon, and it was reported in the news that during a prolonged unusually hot spell, a steel drawbridge that was raised could not be lowered until evening. In each case, the expansion and contraction was caused by changes in the speed with which the material's particles were moving.

There is also evidence that the rate of expansion and contraction is not the same for all materials. Welding two different metals together, for example, makes possible the design of thermostatic controls for use on automatic devices. It is also basic to the design of some unusual "jumping coins" which hold great fascination for students as they engage in an activity that investigates the cause of the jumping action.

Recommended Grade Level: Grades 4–8

Strategies Involved: Hands-on activity
                          Student involvement
                          Student interaction
                          Science skills development

Materials Required:

Each group of four students will need:

- four Jumping Quarters*, one for each student
- a cool, flat surface, such as a table top or metal tray

Procedure:

Instruct students, working in groups of four, to proceed as follows:

1. Apply gentle pressure while rubbing the surface of each of your group's four special quarters, and then lay the coins on the cool, flat surface of a table top or metal tray.
2. Observe the spontaneous jumping action of the coins, and repeat the activity several times. Include some variations, such as:

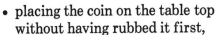

- placing the coin on the table top without having rubbed it first,
- placing a rubbed coin on the table top with the opposite surface of the coin positioned against the table top,
- placing a rubbed coin on a table top that has been covered with paper toweling.

3. Discuss with other members of your group what happens in each of the various conditions included in the activity.
4. Examine the coins closely, making note of the bimetallic construction. Then, prepare a group statement explaining how different rates of expansion by the two different metals that make up these coins caused them to act as they did.
5. As a follow-up, relate the jumping coins to other situations involving bimetallic objects, such as those used in thermostats for home heating systems, electric toasters, refrigerators, and various other devices.

2-14 UNVEILING THE SECRET OF THE DIVING RAISINS

Motivation: An activity for watching objects as they alternately rise and sink in a liquid makes for more than a mere observation. It raises questions in the minds of student observers, who, working in small groups, are encouraged to discuss and share their thoughts and ideas. Some students may be reminded of fish changing their level in a body of water or of the action of a submarine that can dive and resurface by making adjustments in its ballast to change the weight of the vessel—making it heavier or lighter than the volume of water it displaces. By applying scientific knowledge that lends support to the solution of a new problem, students become active participants in the search for an explanation to an attention-getting and puzzling phenomenon.

*Available from Edmund Scientific Co., Barrington, N.J. 08007.

Recommended Grade Level: Grades 5–8

Strategies Involved: Hands-on activity
Student involvement
Student interaction
Science skills development

Materials Required:

Each group of four students will need:

- one small bottle of freshly opened club soda
- 12 raisins
- a tall glass jar

Procedure:

Instruct students, working in groups of four, to proceed as follows:

1. Pour club soda from a freshly opened bottle into a clean glass jar until it is three-fourths filled with club soda.
2. Gently drop 12 raisins, one at a time, into the liquid in the jar.
3. Observe the activity of the raisins placed in the club soda and note any changes in the regularity of the pattern established.
4. Analyze the chain of events observed and discuss with other group members some of the specific factors that are involved:

- evidence of activity before the addition of raisins,
- identification of the bubbles seen rising in the club soda,
- collection of small bubbles on the skin of the raisins coming to rest at the bottom of the glass jar,
- gradual rising of raisins to the surface,
- eventual sinking of raisins to the bottom once again,
- change in rate of activity as diving and surfacing of raisins is repeated over a period of time.

5. Write a short account of "The Secret of the Diving Raisins" that describes how carbon dioxide bubbles collect on the skins of raisins resting at the bottom of the jar and buoy the raisins to the surface, where the bubbles escape, causing the raisins to dive to the bottom once more and repeat the process until all of the carbon dioxide from the carbonated drink has escaped into the atmosphere.

3

Science for Living

One of the foremost and commonly agreed-upon objectives of American education is "to prepare students to function effectively in their roles as informed and responsible citizens in their world of the future." Especially in the field of science, however, this is a difficult goal to achieve because, based on the rate at which scientific progress is being made, it is projected that much of the scientific information of today will be outdated by the time today's middle school/junior high students take their places as leading citizens of the 21st century. In any case, today's students are interested in the *present* and are curious about things to which they can relate *now,* in their *real world.*

Many situations encountered in students' daily lives serve as a source of relevant problems for investigation. The classroom, school grounds, home environment, vacation spots, shopping malls, school bus, sporting and amusement events, TV commercials, and newspaper advertisements, all should be considered. In addition to preplanned problems for study, identification of problems by students should be encouraged to give greater credibility to the investigation of situations that they perceive to be "real."

A variety of motivating strategies can be used in developing activities that focus on SCIENCE FOR LIVING. The learning experiences, involving various aspects of Earth Science, Physical Science, and Life Science, then become meaningful to students because they are real and not contrived and because they relate to student interests.

EARTH SCIENCE ACTIVITIES

3–1 INVESTIGATING A "FAULT" IN THE EARTH'S SURFACE

Motivation: News reports of an earthquake occurring at some location on earth are generally accompanied by dramatic pictures of long cracks in the earth's surface caused by the disturbance, and of the resulting broken windows, toppled buildings, and discontinuous roadways. These news accounts prompt questions. How can something as solid as the earth be made to shake so violently and why do these disturbances occur with greater frequency in some parts of the world than in others? By investigating one of the major causes of earthquakes, students gain some insight into one kind of earth disturbance and some understanding of how a knowledge of the earth's fault lines make possible the prediction of where future earthquakes are likely to occur.

Recommended Grade Level: Grades 6–8

Strategies Involved: Hands-on activity
Student involvement
Student interaction
Science skills development

Materials Required:

Each group of four students will need:

- two pieces of a wood block two-by-four that can be fitted together along a smooth, slightly diagonal seam
- Magic Markers or colored pencils in four different colors

Procedure:

Instruct students, working in groups of four, to proceed as follows:

1. Prepare a simulated section of the earth's crust along an identified fault line:

   - Use a ruler to divide and mark the thickness of both pieces of the two-by-four into four layers.
   - Using a different color for each layer, color both pieces of the two-by-four, including the sloping edges of the diagonal cut, so that when fitted together the two pieces represent a continuous section of the earth's crust.
   - Mark the tops of the two pieces for identification:
     #1 on the piece with the downward slope,
     #2 on the piece with the upward slope.

2. On a smooth table top, place the two pieces of the two-by-four together, with their sloping edges touching and fitted together. Note that the matching layers are continuous, as in layers of the earth's crust.

3. Hold piece #1 firmly and push piece #2 toward #1, maintaining pressure and observing the effect of this pressure along the fault line.

4. Discuss with other members of your group the conditions under which layers of the earth's crust may be disturbed and cause an uneven surface at a fault line.

5. Write a group report that describes the use of the model in demonstrating how the shifting of the earth's crust causes an earthquake.

Encourage students to use their models to investigate disturbances of the earth's crust along a fault line when pressure is exerted in other directions, such as:

- horizontally from both ends,
- horizontally from both sides,
- in an upward or downward direction on one side,
- in an upward or downward direction on both sides.

3–2 INVESTIGATING FOLDED MOUNTAINS

Motivation: Major highway construction often cuts through mountains, revealing a vertical view of the rock layers all the way from the top of the mountain to road level. In many regions this cut surface resembles a series of large ocean waves and attracts considerable attention. An activity that investigates the forces that cause multiple rock layers to be crumpled into a series of folds relates to the processes involved in the formation of many mountain systems throughout the United States and abroad.

Recommended Grade Level: Grades 7–8

Strategies Involved: Hands-on activity
Student involvement
Science skills development

Materials Required:

Each group of three students will need:

- several sheets of construction paper in three different colors
- access to classroom and library reference materials

Procedure:

Instruct students, working in groups of three, to proceed as follows:

1. Allow each student to select construction paper in a favorite color so that three different colors are represented in each group.

2. After each student arranges several thicknesses of paper in his/her chosen color, organize a vertical stack comprised of three horizontal layers, each a different color.

3. Check to make certain that all layers are aligned to form straight edges. Then, hold the stack firmly by placing one hand at each end.
4. Apply pressure at both ends by pushing toward the middle of the stack.
5. Observe what happens to the layers of colored paper and discuss with other group members:

   - the position now taken by the layers that were originally in a horizontal position,
   - the reasons why the layers were rearranged to form an upward bulge.

6. Then relate the activity to some observations of actual earth formations:

   - Cite evidence to indicate that the earth's crust is made up of several layers.
   - Describe the appearance of the rock layers that can be viewed where highway construction has cut through a mountain.
   - Suggest a reasonable explanation for the formation of these layers.
   - Relate the squeezing together and buckling of rock layers, due to pressure which builds up when tectonic plates collide, to the formation of "folded" mountains.
   - Indicate some of the things that folded mountains can tell us about their formation.

7. Use library and classroom reference materials to research examples of folded mountains throughout the world. Then, write a group statement that tells why the Pocono, Appalachian, and many other well-known mountains, such as the Himalayas and the Urals, are called folded mountains.

PHYSICAL SCIENCE ACTIVITIES

3–3 CONTROLLING CHEMICAL CHANGES THAT CAUSE PROPERTY DAMAGE

Motivation: It is customary to apply, periodically, a fresh coat of paint to houses and fences made of wood and to bridges and guardrails made of iron and steel. Of even greater importance than the improved appearance is the protective maintenance feature, which offers protection against damage due to decay and corrosion. The manner in which this practice operates to control such a chemical change can be investigated in an activity that relates to many student experiences.

Recommended Grade Level: Grades 4–6

Strategies Involved: Hands-on activity
Student involvement
Science skills development

Materials Required:

Each group of four students will need:

- two nails with large heads
- two glass jars with flat bases
- an oil-base paint
- a small paint brush
- water
- paper toweling or other protective material
- a metric ruler

Procedure:

Instruct students, working in groups of four, to proceed as follows:

1. Prepare a work area in which table tops and floor are protected by a covering of paper toweling or newspaper.
2. Paint one nail with an oil-base paint.
3. When the paint has dried thoroughly, prepare two jars, labeled #1 and #2:

   - Place the painted nail, standing on its head, in jar #1.
   - Place the unpainted nail, standing on its head, in jar #2.
   - Without disturbing the upright position of the nails, carefully add water to both jars to a depth of 2.5 cm.

4. Without disturbing the position of the nails, place the two jars on a shelf or table top where they can be observed on a daily basis.
5. Be alert to changes that occur and report to other group members any evidence of discoloration of the nails or the bottom of the jars upon which they rest.
6. Discuss with other group members:

   - the appearance of a newly formed substance,
   - the conditions that favored, and those that inhibited, the formation of this new compound,
   - the chemical composition of this new compound,
   - the chemical change involved in rust formation.

7. Relate the activity to some practical situations, such as:

   - a coating of tin over another metal to make a "tin" can,
   - the coating of oil a workman places on his metal tools,

- the oiling of a metal hinge,
- the periodic painting of steel bridges,
- the association of water with the process of rust formation.

8. Then, prepare a group statement that tells how the practice of painting, oiling, and/or coating various surfaces can reduce the estimated millions of dollars that are wasted each year because of the rusting of iron and steel.

3–4   DESIGNING A TUNNEL SO THAT THE ROOF WON'T CAVE IN

Motivation: Although tunnel construction is a costly and time-consuming activity, the expenditure can be justified by the savings in time and energy provided by a travel route that goes through, rather than over or around, a mountain or body of water. However, safety must also enter into the construction plans so that the tremendous weight of the material above the tunnel does not cause the roof to cave in. An activity in which students work in teams assigned to investigate the relative safety of tunnels of different construction design provides for an interesting science lesson.

Recommended Grade Level: Grades 6–8

Strategies Involved: Hands-on activity
Student involvement
Inquiry/discovery approach
Science skills development

Materials Required:

For each eight students, forming two teams of four students each, the following will be needed:

- two cardboard tubes from rolls of paper towels (about 28 centimeters in length)
- two cut-down plastic 1-gallon jugs
- two flashlights
- two tall jars, milk cartons, or empty juice cans
- two measuring cups
- two metric rulers
- two X-acto knives

> NOTE: STUDENTS MUST BE CAREFULLY SUPERVISED
> WHEN USING SHARP INSTRUMENTS.

- sand
- water
- plastic packaging tape

Procedure:

Instruct each group of eight students to form two teams (A and B) and, working with identical cardboard tubes, to proceed as follows:

1. Begin the activity:

   - Team A, reinforce the walls of the tube by wrapping strong plastic packaging tape around the tube so that the outside surface is completely covered.
   - Team B, do not reinforce the tube in any way.

2. Proceed with the activity, with both teams following the same directions:

   1) At a distance of 2.5 cm from the bottom of a cut-down plastic 1-gallon jug, trace the outline of the open end of the tube on the outside wall of the jug.
   2) Using the X-acto knife and following the circular outline of the tube, cut a hole in the side of the jug.
   3) Insert the tube through the circular hole and push it through to the opposite side of the jug.
   4) In a similar manner, mark an outline on the opposite side of the jug and cut an opening that will allow the tube to pass completely through the jug. Then position the tube so that its extension beyond the jug is the same at both ends.
   5) Fill the jug with sand, allowing it to fill in below and around, as well as above the tube.

   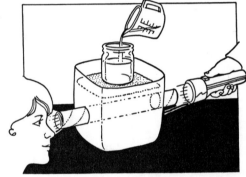

   6) Select one member of the team to hold a flashlight at one end of the tube while other members, at the opposite end, view and draw a diagram of the beam of light that passes through the tunnel.
   7) Place a large jar on top of the sand and, while one team member pours a cupful of water into the jar, others can view the light through the tunnel to determine its shape and detect any changes that occur as additional cupfuls of water are added.
   8) Determine how much water was added before the weight caused the tunnel to show signs of caving in, as observed by differences in the shape of the light through the tunnel.

3. Participate in a group discussion in which teams A and B compare experiences, noting any differences and giving attention to:

   - the amount of weight that was withstood before any change was noted,
   - the relative strength of reinforced and unreinforced tunnel walls.

4. Report to other members of your group any personal experiences and observations made while traveling through a tunnel and suggest:

   - other methods of reinforcement, such as supports along the inner walls,
   - reasons why greater reinforcement of tunnel walls is needed in tunnels running through limestone formations than in those passing through an area of hard rock.

3–5 UNDERSTANDING REFRIGERATION SYSTEMS

Motivation: Insulated ice chests are commonly used to keep foods cool for picnics and other outdoor events away from home, but the standard type of refrigeration appliance found in most homes today passes a refrigerant material through a system of cooling coils. The same gas that acts as a refrigerant here is also commonly used in pressurized cans of commercial products, such as insect repellents, air fresheners, deodorants, and hair sprays. It is possible to use one of these products to investigate the principle of the refrigeration process.

Recommended Grade Level: Grades 6–8

Strategies Involved: Hands-on activity
Student involvement
Manipulation of laboratory materials
Science skills development

Materials Required:

Each student/partner team will need:

- one can of air freshener or other product in a pressurized can
- a length of copper tubing
- water

Procedure:

Instruct students, working with partners, to proceed as follows:

1. While your partner holds one hand wrapped around a piece of copper tubing, direct the outlet of the spray mechanism of a can of air freshener (or other product in a pressurized can) toward the inside of one of the open ends of the tube.

2. After your partner reports what happened as the spray was discharged into the tube, exchange places with your partner so that you, too, can experience the sensation.
3. Repeat the process, this time coating the outside of the tube with water before releasing the spray and observe what happens to the water. Relate the formation of frost, and the sensation of cold on the hand, to the transfer of heat as it passed through the metal walls of the tube and caused the spray within to evaporate.
4. Then, discuss with your partner the basic scientific principle involved in a refrigeration system by tracing the main events:

- A liquid refrigerant is moved through a closed system of coils.
- Heat from the refrigerator compartment flows into the tubing and causes the refrigerant to vaporize.

- A condenser outside the refrigerator causes the vapor to condense back to a liquid form.
- The cycle is repeated, removing heat energy and keeping the interior of the refrigerator cold.

5. Make a list of appliances, other than refrigerators, that are based on this principle and tell why a refrigerator will not work if the refrigerant gas escapes and is lost.

3–6 INVESTIGATING HOW AIR IS USED TO LIFT THINGS

Motivation: A visit to an auto service center will provide an opportunity to observe the seemingly effortless manner in which a car is raised to a height that permits the mechanic to make needed repairs on the underside of the car. Actually, an increase in air pressure is needed to exert enough force to cause the mechanism on which the car is supported to rise to the desired height. How air is used to lift things can be investigated by engaging in a simple activity.

Recommended Grade Level: Grades 4–6

Strategies Involved: Hands-on activity
Student involvement
Science skills development

Materials Required:

Each student/partner team will need:

- two styrofoam coffee cups of the same size

Procedure:

Instruct students, working with partners, to proceed as follows:

1. Taking turns, so that one student can observe while the other performs the activity:

   - Place one styrofoam cup inside the other so that one cup is nested within the other.
   - Take a deep breath and then blow a steady, long, hard breath on the side of the top cup where it rests against the lower cup.

2. Observe what happens as each student performs the activity.
3. Discuss with your partner:

   - how blowing air into the space between the side walls of the two cups reduced the air pressure between the sides of the cups and caused a greater volume of air to be trapped beneath the top cup,

- how the pressure, exerted by the air trapped beneath the top cup, caused it to rise to a higher level.

4. Suggest ways to maintain the amount of pressure that is needed to lift the top cup so that the effect of lifting can be prolonged.
5. Prepare a list of devices that make use of air pressure to lift things—and invent a few new ones as well.

3–7   ANALYZING THE FORMATION OF A CHEMICAL COMPOUND

Motivation: Silver tableware that comes in contact with certain sulfur-containing foods becomes blackened by the formation of a new compound called silver sulfide. Generally, the film that forms on silver utensils used for eggs, mustard, and mushrooms can then be removed by polishing the affected pieces of silver. Tarnished silver can also be cleaned by a chemical reaction in which the sulfur is caused to combine with another metal, such as aluminum, thereby dissociating itself from the silver.

Recommended Grade Level: Grades 6–8

Strategies Involved: Hands-on activity
Student involvement
Inquiry/discovery approach
Science skills development

Materials Required:

Each group of four students will need:

- tarnished silverware that has been thoroughly washed
- aluminum foil
- an enamel pan
- a hot plate
- tongs
- measuring spoons
- paper toweling
- table salt (sodium chloride, NaCl)
- baking soda (sodium bicarbonate, $NaHCO_3$)
- water
- a jar of mustard

Procedure:

Instruct students, working in groups of four, to proceed as follows:

1. Set up a work station that is equipped for heating solutions.
2. Obtain a few pieces of tarnished silverware. (These can be produced quickly by dipping silver spoons in a jar of mustard for a period of about 10 minutes or until a dull film of silver tarnish (silver sulfide, Ag_2S) covers their surface.)

3. Place a 6-inch square of aluminum foil in the bottom of an enamel pan that is three-fourths filled with hot water.

4. Using the capacity of the pan as a guideline for determining the volume of water in the pan, add 1 teaspoon of salt and 1 teaspoon of baking soda for each quart of water in the pan.

5. TAKING CARE TO PRACTICE PROPER SAFETY PRECAUTIONS, place the pan on a hot plate and heat the solution in the pan. Then remove the pan from the heat source and allow it to cool slightly.

6. Using tongs, gently immerse some of the tarnished silver pieces, one at a time, into the solution, retaining some pieces to be left untreated.

7. After three or four minutes, use the tongs to remove the treated silver pieces from the solution and set them on toweling.

8. Examine the appearance of the square of aluminum foil and the silverware that has been treated. Then discuss with members of your group:

   • the transfer of the dull finish from the tarnished silverware to the aluminum foil,
   • the expected results if an aluminum pan were to be used instead of an enamel pan.

9. Compare the treated silverware and the untreated tarnished pieces, and consider:

   • what caused the silverware to become tarnished,
   • the chemical composition of silver tarnish,
   • whether metals other than silver can become tarnished.

10. Then, consider the practical applications of the scientific principle that was involved in this activity:

   • What are some sources of sulfur that cause silverware to become tarnished?
   • Why is the method of tarnish removal illustrated in this activity called "chemical displacement?"
   • Why do some people prefer stainless steel tableware to silver, and stainless steel cookware to aluminum pots and pans?

3–8   CHECKING THE EFFECTIVENESS OF AN "ANTIFREEZE"

Motivation: Heat that is generated by most motors causes overheating, which interferes with the efficiency of the motors and their operation. The use of a coolant prevents overheating, as can be observed in the cooling system of an automobile, where water in a radiator circulates to keep the engine cool. However, a problem arises during winter months when the temperature drops below the freezing point of water. The problem can be solved by the addition of certain materials to the water to produce an "antifreeze" mixture.

Recommended Grade Level: Grades 4–6

Strategies Involved: Hands-on activity
                         Student involvement
                         Science skills development

Materials Required:

Each group of four students will need:

- two identical, small, empty juice cans, with tops removed and with no jagged edges
- water
- a 50 percent solution of rubbing alcohol in water
- a colored wax pencil or other marker
- a thermometer
- a ruler
- access to a freezing compartment

Procedure:

Instruct students, working in groups of four, to proceed as follows:

1. Measure the height of two identical juice cans and make a mark on the side of each indicating the two-thirds full level. Then, label the cans #1 and #2, respectively.
2. Fill can #1 to the two-thirds full level with water and fill can #2 to the two-thirds full level with the 50 percent solution of alcohol in water.

   CAUTION: AVOID INHALING THE FUMES OF THE RUBBING ALCOHOL OR PERMITTING IT TO COME IN CONTACT WITH THE HANDS OR ANY OTHER PART OF THE BODY.

3. Place a thermometer and both cans in the freezing compartment of a refrigerator and allow them to remain, undisturbed, overnight.
4. The next day, take a temperature reading of the freezer compartment and remove the cans from the compartment for a comparison of their contents.
5. Work with other group members to determine:

   - the appearance of the contents of cans #1 and #2,
   - the temperature of the freezing compartment,
   - the effectiveness of alcohol, at a 50 percent concentration, as an antifreeze agent at the temperature of the freezing compartment.

6. Then, discuss some additional considerations:

   - Is there any temperature at which an "antifreeze" substance will freeze?
   - What effect would the concentration of alcohol in water be expected to have on the freezing point?
   - How would a knowledge of the scientific principle involved in antifreeze preparations be helpful to a person who lives in an area where a temperature drop to 20 degrees below 0° F. has been predicted?
   - What other substances are used as antifreeze agents? What are the advantages and disadvantages associated with each of these agents?

7. With other members of your group, prepare a good definition of the term "antifreeze."

LIFE SCIENCE ACTIVITIES

3–9   SERVING FOOD AT SALAD BARS

Motivation: Salad bars have become very popular in restaurants, grocery markets, and fast-food shops, while fresh fruit and vegetable platters are often used for daylong snacks at picnics and barbecues. This poses a problem for all food handlers, whether at home or in the food industry. The familiar sight of an apple turning brown within minutes of being sliced is but an example of the problem. How to overcome this problem in an efficient way that does not alter the food or introduce some potentially harmful additive or preservative can be investigated in the classroom.

Recommended Grade Level: Grades 4–8

Strategies Involved: Hands-on activity
Student involvement
Science skills development

Materials Required:

Each group of four students will need:

- an apple
- a knife
- a saucer
- lemon juice in a small cup
- a timing device

Procedure:

Instruct students, working in groups of four, to proceed as follows:

1. WHILE PRACTICING EXTREME CAUTION, use a knife to cut an apple in half.
2. Dip the cut surface of one half-apple in lemon juice.

3. Place both halves, treated and
   untreated, cut surface up, on a
   saucer.
4. Observe the appearance of the
   two surfaces and record the
   time that elapses before each
   begins to turn brown.
5. Discuss with members of the
   group the differences noted
   in step 4 and the reasons for
   these differences:

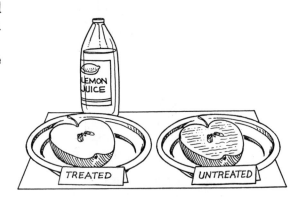

- the browning of cut apple surfaces exposed to air as caused by the action
  of the enzyme polyphenolase—naturally present in the living tissue of
  many fresh fruits and vegetables,
- the effective protection against browning offered by lemon juice, which
  slows the chemical reaction between a freshly cut apple and oxygen in the
  air by inactivating this enzyme, normally found in the living tissue.

6. Suggest ways in which a knowledge of the science involved in this activity is
   useful in life experiences:

- Name other foods that could be "preserved" in a similar manner.
- Tell why the same problem does not exist if the fruits and vegetables are
  cooked before being cut and exposed to the air.
- Indicate possible advantages of using lemon juice rather than a chemical
  preservative to prevent food spoilage due to oxidation.

3–10   CHECKING THE NEED TO "REFRIGERATE FOODS AFTER OPENING"

Motivation: Bulging cans are an indication of the presence of harmful microorganisms in some canned foods—the bulging being due to the accumulation of carbon dioxide produced by the organisms metabolizing the foodstuffs within.
However, because only a few species of microorganisms can live in the oxygen-free
environment provided by a sealed can or bottle, most foods are protected from such
an invasion. After being opened, of course, there is a real danger that these foods,
too, can become contaminated by the entrance of air and microorganisms from the
outside. If and when this occurs, steps must be taken to prohibit the microorganisms from engaging in their normal life activities, which would lead to the eventual
spoilage of the food.

Recommended Grade Level: Grades 4–6

Strategies Involved: Hands-on activity
                         Student involvement
                         Science skills development

Materials Required:

Each group of four students will need:

- two unopened glass bottles (with screw-cap tops) of cranberry juice cocktail
- two labels
- four paper cups
- access to a refrigerator
- a marking pen or pencil

Procedure:

Instruct students, working in groups of four, to proceed as follows:

1. Observe the appearance of the cranberry juice cocktail in two unopened bottles, noting the sparkling-clear liquid through the glass walls of the containers.

2. Open the containers in the usual way and, from each bottle, remove two paper cupfuls of juice to be enjoyed by members of the group as a mid-morning or mid-afternoon refreshment.

3. While the bottles remain open, prepare and affix labels that will identify the bottles as belonging to your group and distinguish between the two bottles, #1 and #2.

4. Replace the screw-cap tops on the bottles. Then place bottle #1 in a refrigerator and bottle #2 on a table or open shelf.

5. Each day examine the contents of both bottles and record all observable differences noted in their appearance.

6. Continue to collect data for a period of one school week. Then, based on the data collected, discuss with other members of the group:

   - any observable evidence of microorganisms living in the juice in either bottle,
   - the probable source of the microorganisms in the bottle(s),
   - conditions that appear to favor the growth of microorganisms,
   - conditions that appear to inhibit the growth of microorganisms.

7. Compare the results of the investigation with those obtained by other groups engaging in the same activity.

8. Prepare a group statement that explains why products such as cranberry juice cocktail can remain on open, unrefrigerated shelves in grocery stores and pantries *before,* but not after being opened.

9. Read the warning on the product label and suggest several reasons why heeding this warning is a good practice.

10. As a follow-up, collect labels from other products that supply warnings about contamination of food by microorganisms whose presence is accompanied by conditions that favor their life activities.

3–11 TRACKING TEMPERATURE LEVELS NEEDED FOR LIFE ACTIVITIES

Motivation: We do not usually envision our being able to live on other planets, such as Pluto, Saturn, Jupiter, and Mars. Among the unfavorable conditions for life as we know it, extremes of temperature are known to exist on other planets, thereby ruling out the possibility of certain chemical reactions that occur only at specified moderate temperatures. Since many of these reactions must occur within each organism as it engages in life activities, life, as we know it, is limited to a very narrow range of temperatures. Some organisms, such as birds and mammals, have a built-in control mechanism that maintains a relatively constant body temperature, but a far greater number have body temperatures that fluctuate with the temperature conditions of their surroundings. This sets restrictions on the chemical reactions and related life activities of variable body temperature organisms, making them suitable subjects for investigations into the effect of temperature on some biochemically controlled activities of organisms.

Recommended Grade Level: Grades 6–8

Strategies Involved: Hands-on activity
Student involvement
Inquiry/discovery approach
Science skills development

Materials Required:

Each group of four students will need:

- active houseflies
- two clean mayonnaise jars with one-hole rubber stoppers to fit
- two thermometers
- two low bowls or pans
- two labels and/or a marking pen
- crushed ice

Procedure:

Instruct students, working in groups of four, to proceed as follows:

1. Prepare two identical fly habitats. For each:

   - Moisten a thermometer and carefully insert it through the hole in a rubber stopper.
   - Place four or five captured houseflies in a mayonnaise jar and quickly position the rubber stopper in the mouth of the jar.

- Adjust the level of the thermometer so that it extends halfway into the jar.
- Set the assembled mayonnaise jar in a low bowl on a table top.
- Allow the flies to adjust to their new surroundings. Then, make note of their activity and take a temperature reading.

2. Attach labels to the bowls, identifying them as Bowl A and Bowl B. Then, place some crushed ice in Bowl A, packing it around the lower end of the jar that is positioned in it.

3. Observe the two setups for several minutes and make note of:

- any temperature changes that occur in jars A and B,
- any changes in the activities of the flies in habitats A and B.

4. Discuss with other members of the group:

- the relationship between temperature and the observed activity of the flies,
- how flies that have been rendered inactive might be reactivated,
- the reason for including setup B, which was not designed to involve a temperature change.

5. On a volunteer basis, design and conduct a follow-up activity that checks out whether the same results would have occurred if some air holes had been provided for the flies in the jars.

6. Suggest possible explanations for some related situations:

- Houseflies are more commonly seen in summer than in winter months.
- A mouse, a canary, or a human remains active at temperature levels observed to deactivate flies.

3–12 COMPARING THE DIGESTIBILITY OF BREAD AND TOAST

Motivation: Food and nourishment are required to supply the energy needed for sustaining human life as well as life in lower forms. However, it is not uncommon for sick people to be unable to enjoy their favorite foods or to eat normal meals. During such times Dr. Mom usually prescribes a light diet of tea and toast to help sick persons keep their strength, which helps them to recover from their illness. A knowledge of the subtle chemical change that occurs when plain bread is toasted is important for developing an understanding of how a sick person's system can tolerate toast, but not the bread from which it is made.

Recommended Grade Level: Grades 5–8

Strategies Involved: Hands-on activity
Student involvement
Science skills development

Materials Required:

Each group of four students will need:

- small strips of white bread
- small strips of toasted white bread or melba toast
- Lugol's iodine solution in a dropper bottle
- a small saucer
- paper toweling

Procedure:

Instruct students, working in groups of four, to proceed as follows:

1. Protect a work area by covering a table top with paper toweling.
2. Place a small bit of white bread and a small bit of the browned portion of a piece of toasted bread in well-separated positions on a saucer.
3. PRACTICING EXTREME CAUTION IN THE USE OF THE IODINE SO-LUTION*, add a drop of weak iodine solution to each sample.
4. Observe the color changes that occur when iodine solution is added to plain bread and to toasted bread.
5. Discuss with other members of the group:

- the typical blue-black color produced where an iodine/starch complex was formed with plain bread and the reddish brown color that appeared where iodine solution came in contact with toast,
- the partial hydrolysis (digestion) of starch that occurs when bread is toasted,
- how the conversion of starch to dextrin when bread is toasted makes it easier for us to digest toast than plain bread,
- the anticipated reaction of iodine if added to bread that has been fully digested.

6. Recall the taste of toast (dextrin) and compare it with the remembered taste of plain white bread. Then account for the difference in taste.
7. Write a group statement relating the initial steps in digestion of starch to the heating of white bread when it is toasted. Then, tell how toasting makes the bread easier for sick people to digest.

*NOTE: Iodine will stain clothing, skin, and/or other objects with which it comes in contact. Prompt rinsing of an affected area with clear water will minimize the intensity of the stain.

4

Using
the Process Approach
in Science

Although not all will become scientists, students in the middle grades derive great satisfaction from engaging in investigations in which they "work as scientists." Through these experiences they develop skills that can be used when seeking answers to many questions and when searching for the solutions to many problems.

The study of science begins with the *natural* curiosity of students, but it is investigative activities, *designed* to help students develop a progressively more sophisticated approach to discovering the process of science, that are the most effective for use in the middle school science program. The goal is to guide students toward the development of higher level thinking skills, while strengthening their manipulatory skills through active involvement in a variety of challenging hands-on activities. The expected outcome is that students will discover for themselves the meaning of science, while at the same time increasing their self-confidence and their ability to use scientific skills and attitudes in problem-solving situations.

The investigative approach recognizes that students must play the major role in their learning if true understanding is to be achieved, and that the teacher interacts with students only when needed to promote the process of student inquiry. It allows students great freedom—to reflect on their ideas, to interpret observations made, to exchange ideas with other members of their group, to analyze the data collected, to apply pertinent information to personal experiences, and to extend their thoughts and "discoveries" to new situations. After completing a student-centered activity, the entire class can then be brought together for a debriefing session in which the teacher helps the students to refine their thinking about the investigation. Student input can be encouraged by inviting their responses to questions, such as:

- "What did you find out?"
- "What data collected supports this 'discovery'?"
- "Would this information be useful in answering a different question in another situation?"
- "Could the method used in your investigation be useful in a different situation?"
- "Are there any questions left unanswered?"
- "Can you suggest a method of finding answers to those unanswered questions?"

All levels of learning are emphasized in this approach. Students become involved in activities that focus on areas of scientific knowledge, comprehension, application, analysis, synthesis, and evaluation.

ACTIVITIES FOR DEVELOPING SCIENCE PROCESS SKILLS

4-1   OBSERVING WHAT HAPPENS WHEN WATER BOILS

Motivation: It is easy to determine when water boils—the water becomes extremely hot and observations can be made of bubbles forming within the liquid and of steam escaping into the atmosphere. Careful attention to details, observed at close range, will provide students with an opportunity to describe the events that comprise the boiling process—thus sharpening their observational skills and indicating their understanding of the scientific knowledge gained from the activity.

Recommended Grade Level: Grades 6–8

Strategies Involved: Student involvement
Science process skills development
Hands-on activity

Materials Required:

Each group of four students engaging in the activity will need:

- a heatproof, transparent heating vessel
- a hot plate
- water
- safety goggles
- small beads

Procedure:

CAUTION: ALL STUDENTS MUST WEAR SAFETY GOGGLES AND KEEP AT A SAFE DISTANCE FROM THE HEAT SOURCE AND BOILING WATER AS THEY ENGAGE IN THIS ACTIVITY.

Instruct all students, working in groups of four, to proceed as follows:

1. Fill a heatproof beaker or glass cooking vessel about half full of water.
2. Place the beaker on a hot plate, or other heating source, set at a moderately high heat level.
3. Observe the activity as the water is heated to the point of boiling, and record all changes that are observed to occur.
4. Discuss with other group members those things that were observed to happen when water boils, noting:

- the escape of steam into the air above the container,
- the formation of small bubbles near the bottom of the container,
- the increase in size of the small bubbles, which eventually rise to the surface and escape into the air,
- the decrease in volume of water as it boils.

5. Relate the bubbling action observed to the increased energy level of the heated water molecules, which no longer allows them to remain quiet and close to each other:

- escape of bubbles at the surface,
- accumulation of energy as small bubbles at the bottom grow and push their way to the surface.

6. Extend the activity to include the addition of small beads in the bottom of the beaker and observe the effects the beads have on:

- the size of bubbles formed at the bottom,
- the accumulation of energy, which forms large bubbles,
- the rate at which bubbles at the bottom rise to the top and escape in the form of steam.

7. Write a group statement that describes what must happen to the particles that make up water to cause them to become a gas.

4-2   DETERMINING THE NEED FOR MEASURING DEVICES

Motivation: Because our senses often misrepresent the messages they receive, observations based on information gained by the senses alone may not always be reliable. This can be demonstrated effectively by the use of certain diagrams that trick our senses into perceiving the relative size of two objects in one way, while actual measurements reveal just the opposite. The use of precise measurements should be employed whenever possible for collecting unbiased scientific data; it rules out the danger of distortions and incorrect interpretations of information gathered by methods that depend largely on impressions.

Recommended Grade Level: Grades 4–8

Strategies Involved: Hands-on activity
Student involvement
Science process skills development
Student interaction

Materials Required:

Each student engaging in the activity will need:

- a copy of the top hat "optical illusion"
- a ruler

Procedure:

Instruct students to proceed as follows:

1. View the "illusion" diagram and, based on observations made, respond to the question—"Is the hat taller than its brim is wide?"

2. Report your response to the class secretary, who will record the responses of all class members in the form of a tally on the chalkboard.

3. Then, using a ruler, measure the height of the hat and the width of its brim, expressing the measurements in units that have been stated for use by all members of the class.

4. Compare your response made as a result of a visual observation with one that can now be made, based on measurement.

5. Refer to the tally on the chalkboard and discuss with other members of the class:

   - the class record for correct and incorrect responses, made on the basis of observation alone,
   - some possible reasons why the optical illusion misled some students into believing that the height of the hat was greater than the width of its brim,
   - why some observations, if made with the senses alone, may lead to wrong conclusions.

6. Write a short account of the activity, telling why, in scientific work, it is best to use precise measurements whenever possible.

As an extension of the activity, provide other "illusions" for students to consider and ask them to indicate appropriate measuring devices that would be useful for making an accurate measurement in each of the situations.

Is the hat taller
than its brim is wide ?

4–3 SELECTING AN APPLE A DAY

Motivation: "An apple a day" is an expression that could easily apply to many school lunches, which often include a piece of this tasty fruit. In addition to good taste and high nutritional value, an apple is a good source of natural fiber, vitamins, and minerals. The many kinds of apples that are available accommodate individual preferences in matters of sweetness, juiciness, and firmness of texture, and provide for a different variety to be represented in each day's lunch during the school week. An activity that focuses on external features of specimens, representing different varieties in a collection of apples, sharpens student awareness of the need for attention to observational details to be used in describing an object, including one's favorite kind of apple.

Recommended Grade Level: Grades 4–8

Strategies Involved: Hands-on activity
Student involvement
Science process skills development
Group interaction

Materials Required:

Each group of five students will need:

- a tray on which have been placed five apples (one of each variety: Red Delicious, McIntosh, Golden Delicious, Winesap, and Granny Smith), each flying a toothpick-supported flag that designates a different day of the school week—Monday through Friday.
- an envelope containing a duplicate set of five flags, one for each day of the school week—Monday through Friday.

Procedure:

Instruct students, working in groups of five, to proceed as follows:

1. Observe the apples on the tray and note that each apple represents a different variety and flies a flag for a different day of the school week.
2. Allow each student in the group to select a flag from the envelope and, without revealing to other students which day was picked or disturbing the position of the apples on the tray, to:

    - focus attention on the apple that is flying a flag for the day of the week corresponding to that selected from the envelope,
    - make a list of all observable features of the appropriate apple.

3. Arrange for all members of the group, each in his/her turn, to read the lists prepared while other members of the group try to identify each day's apple on the basis of the description presented.

4. Tally how many descriptions were accurate enough to provide for the proper identification of the apples by members of your group, and report this information to the entire class for a discussion of:

- information found to be most helpful in making a correct identification,
- the importance of making careful observations, both for writing a description and for making an identification of an object that has been described.

Suggest that each student claim his/her "apple of the day" to be enjoyed at lunchtime or as an after-school snack.

4–4 MEASURING THE DEGREE OF ACIDITY OF COMMON SUBSTANCES

Motivation: Although strong acids can cause burns on the skin and eyes, there are many acidic materials in and around the home and classroom that are safe to handle. Students can make a collection of safe, everyday materials and bring them into the classroom to be tested with pH paper. They can then organize their results on a pH scale, listing the materials according to their relative degree of acidity.

Recommended Grade Level: Grades 5–8

Strategies Involved: Hands-on activity
Student involvement
Science process skills development

Materials Required:

Each group of four students will need:

- six baby food jars, each containing a different substance brought in by students (lemon juice, vinegar, milk, club soda, rainwater, aquarium water, or some other common substance)
- six medicine droppers
- six strips of pH test paper
- a pH color chart
- a small glass or porcelain dish or saucer
- paper toweling
- a wax marking pencil

Procedure:

Instruct students, working in groups of four, to proceed as follows:

1. Make some advance preparations:

- Protect the desk top or other work area by covering it with paper toweling.
- Label each jar brought in by group members to indicate the contents of the jar.
- Assemble all sample jars, pH test paper strips, medicine droppers, and small saucer or spot plate to be used in the activity.
- Prepare a chart that provides for recording the name of each substance to be tested and the corresponding pH value, as indicated by the test.

2. Test, individually, each sample of material brought in by group members:

   - Use a medicine dropper to transfer one drop of material from a sample to the end of a small strip of pH test paper being held over a saucer or spot plate.
   - Allow the paper strip to dry and observe it for evidence of a color change.
   - Match the resulting color of the dry test strip with the pH scale and determine its pH number.

3. Record, on the prepared data chart, the name and pH value of the material represented by each sample tested.

4. Discuss with other group members:

   - the variety of pH values represented by the common substances tested,
   - substances for which the pH values were high, as well as those that tested "low,"
   - remembered taste sensations produced by some of the substances tested that suggest a relationship between taste and pH value.

5. Engage in a debriefing session in which all students summarize their findings about the nature and usefulness of pH testing. Then prepare a thermometer-type pH scale that lists all pH numbers in a vertical column and write the name of each substance tested at its proper location on the scale.

4–5 DETERMINING THE MAGNIFYING POWER OF A LENS

Motivation: Some magnifying lenses are marked to indicate their magnifying power. For others that are unmarked, or for determining the magnifying power of a drop of water or a "magic viewer" such as that used in Activity 2-11, students may engage in an activity that allows them to interpret their findings and make an estimate of the magnifying power of an enlarging lens.

Recommended Grade Level: Grades 6–8

Strategies Involved: Hands-on activity
Student involvement
Science process skills development

Materials Required:

Each student/partner team will need:

- lined notebook paper or graph paper
- a pencil
- a ruler
- an unmarked magnifier such as the magnifier top for a small (1 by 1 in.) specimen "Bug" box*

*Available through SCIENCE KIT, Tonawanda, New York, 14150.

Procedure:

Instruct students, working with partners, to proceed as follows:

1. Using a ruler to make precise measurements, divide the space between two consecutive lines on a sheet of notebook paper into four equally spaced divisions. Then draw parallel lines that separate the original space into four narrow spaces of equal width.
2. In a similar manner, subdivide one or two additional spaces on the lined paper until the closely spaced parallel lines span at least 1 inch on the paper.
3. Place the magnifier top of the small specimen box on the paper and count the number of closely spaced lines that fit from one edge of the box to the other.
4. Raise the magnifier to the level at which the lines, when viewed through the lens top, come into sharp focus and count the number of lines on the paper than can be observed.
5. Compare the number of lines that can be seen through the lens when in focus with the number of lines that physically fit under the magnifier box top, and make an inference about what causes this difference.
6. Make a determination of the magnifying power of your box top magnifier.
7. Explain why a lens that spans 15 lines, but allows the viewing of only five lines when in focus position, is said to have a magnifying power of three.

4–6 COMPARING THE TIME REQUIRED FOR DISSOLVING TWO ANTACIDS

Motivation: There are many alkalizing preparations available for treating conditions of excess acid in the stomach and providing relief for sufferers of temporary stomach upsets and disorders. Some are in the form of tablets that can be swallowed with a glass of water, while others, either in tablet or granular form, must be dissolved in water before being taken internally. Checking the time required for some standard dosages of available preparations to dissolve in water may lead to some interesting comparisons of the "fast-acting" nature of some nationally advertised brands of antacids. It also prompts students to make use of some basic scientific knowledge for predicting the effect that different conditions will have on the rate at which such substances become dissolved in water.

Recommended Grade Level: Grades 6–8

Strategies Involved: Hands-on activity
Student involvement
Science process skills development

Materials Required:

Each group of four students will need:

- two drinking glasses or clear plastic drinking cups
- a marking pen or pencil
- one tablet of Alka Seltzer™
- one teaspoon of Bromo Seltzer™
- water
- a teaspoon

Procedure:

Instruct students, working in groups of four, to proceed as follows:

1. Label two glasses, #1 and #2, and fill each glass two-thirds full of water that has been allowed to reach room temperature.
2. Allow one student of your group to remove one tablet from a packet of Alka Seltzer™, while a second student measures out 1 teaspoonful of Bromo Seltzer™. (These are standard doses.)
3. Make predictions, both individual and group, as to which substance will dissolve more quickly when placed in a glass of water.
4. Allow a student to place an Alka Seltzer™ tablet in the water in glass #1 at the same time that another student adds the teaspoonful of Bromo Seltzer™ (without stirring) to the water in glass #2.
5. Observe the activity in the two glasses and comment on:

   - the appearance of bubbles rising to the surface in each glass,
   - which antacid dissolved more quickly.

6. Discuss with other members of the group:

   - the comparison between the predictions made and the actual findings as to which substance dissolved more quickly,
   - the scientific basis for the prediction that was checked out by the activity,
   - the relationship that exists between the amount of surface area of a substance and the time required for it to dissolve.

7. Consider other factors that might be expected to have an effect on the amount of time required for dissolving a substance in water. Then, plan and conduct an activity that would check out the accuracy of predictions made when:

   - an Alka Seltzer™ tablet is crushed before being placed in a glass of water,
   - an Alka Seltzer™ tablet is placed in a glass of water that has been heated or cooled.

4–7   DISCOVERING THE PROPERTIES OF AN UNKNOWN SUBSTANCE

Motivation: Not all materials have the same properties. Some can be observed to bounce, while others stretch, and still others spread out and flow over surfaces with which they come in contact. These properties can be discovered by engaging in open-ended investigations in which students observe a variety of materials, including a tennis ball, an elastic band, and a jarful of honey, molasses, or catsup. A new and unusual substance, such as PLASTEE, has, of course, greater potential as an attention-getter. The excitement of making a discovery about this "new" substance is further heightened when students observe that one of its properties can be readily changed, as if by magic.

Recommended Grade Level: Grades 6–8

Strategies Involved: Hands-on activity
Student involvement
Science process skills development
Group interaction

Materials Required:

Each group of four students will need:

- a small cottage cheese or margarine tub containing PLASTEE
- a sheet of wax paper
- a wooden tongue depressor
- paper toweling

For preparing the PLASTEE, the following materials also will be needed:

- powdered cornstarch
- green food coloring
- a large mixing bowl
- a large wooden spoon
- a medicine dropper
- a measuring cup
- water

Procedure:

Involve students in all phases of the activity:

1. Enlist the aid of student volunteers to prepare the PLASTEE prior to classtime:
   - Combine 1 cupful of water and about 6 drops of green food coloring in a large mixing bowl.
   - Slowly add the contents of a 1-pound box of powdered cornstarch, stirring the mixture constantly with a large wooden spoon.
   - Continue stirring the mixture until it flows (when the bowl is tipped slightly), but has the consistency of a solid when it is touched.
   - Divide the mixture, placing a quantity in each of the small cottage cheese containers or margarine tubs to be used by the student groups.

2. During classtime, allow the students, organized into groups of four, to engage in the activity:

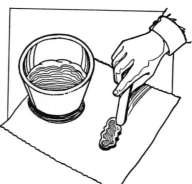

- Protect all exposed surfaces by covering the work area with paper toweling.
- Place a sheet of wax paper on the paper toweling.
- Transfer a small amount of the PLASTEE to the wax paper and examine its appearance.
- Determine the properties of the substance by investigating:

  1) its physical state,
  2) its feel to the touch,
  3) its response to being handled in a variety of ways, such as being probed with a tongue depressor.

NOTE: ALTHOUGH THIS MATERIAL IS COMPLETELY SAFE TO HANDLE, STUDENTS SHOULD BE REMINDED THAT THEY SHOULD FOLLOW AN ESTABLISHED RULE FOR SAFETY IN THE LAB: DO NOT PUT ANY MATERIALS BEING INVESTIGATED IN THE MOUTH.

- Prepare a list of statements based on the group findings about PLASTEE.

3. Invite one member from each group to act as a group spokesperson during a debriefing session:

- Compare individual group findings, with explanations of special techniques used to investigate the properties of PLASTEE.
- Prepare a composite description of PLASTEE.

4. Encourage each student, individually, to engage in one extension of the activity:

- Investigate how the properties of PLASTEE are changed by the addition of a few drops of water.
- Relate research findings about "plasticity" to the properties observed in PLASTEE.
- Suggest one or more practical situations in which PLASTEE could be used.

4–8 PREDICTING pH VALUES OF MIXTURES

Motivation: Lemonade is a refreshing drink which is especially enjoyable on a hot summer day. To suit individual tastes, the basic recipe can be modified to make it strong or weak, as well as tart or sweet. A determination of the pH values of lemon juice and of water will be useful in making predictions concerning the relative pH values of mixtures resulting from the combination of different measured amounts of the two substances.

Recommended Grade Level: Grades 6–8

Strategies Involved: Hands-on activity
                         Student involvement
                         Student interaction
                         Science process skills development

Materials Required:

Each group of four students will need:

- lemon juice
- water
- three baby food jars, of equal size, thoroughly cleaned
- five medicine droppers
- pH test paper
- a pH color chart
- a marking pen or pencil

Procedure:

Instruct students, working in groups of four, to proceed as follows:

1. Assign a number to each of three baby food jars and, with a marking pen, label the jars #1, #2, and #3, respectively.
2. Prepare a chart for recording the predicted and/or measured pH values for the various samples of materials used in the activity.

DETERMINING pH VALUES OF SUBSTANCES ALONE AND IN COMBINATION			
Sample Tested	**Nature of Sample**	**Predicted pH Value**	**Measured pH Value**
Water	Plain water	– – –	
Lemon juice	Undiluted lemon juice	– – –	
Jar #1 Mixture	Water plus 10 drops of lemon juice	– – –	
Jar #2 Mixture	Water plus 50 drops of lemon juice	– – –	
Jar #3 Mixture	Water plus 150 drops of lemon juice	– – –	

3. Determine and record the pH values for basic materials:

   • With a clean medicine dropper, transfer 1 drop of lemon juice to a strip of pH test paper. Then refer to the pH color chart to determine the pH value of the sample and record this value in the proper space on the prepared chart.
   • In a similar manner, measure and record the pH value of the water.

4. By dropper, transfer lemon juice to numbered baby food jars:

   • 10 drops of lemon juice to jar #1,
   • 50 drops of lemon juice to jar #2,
   • 150 drops of lemon juice to jar #3.

5. Add a sufficient amount of water to each jar so that each will become filled with a well-mixed combination of lemon juice and water.

6. With a clean medicine dropper, transfer 1 drop of the "lemonade" mixture in jar #1 to a strip of pH test paper. Then measure its pH value and record this value in the appropriate space on the prepared chart.

7. On the basis of the data collected, predict the pH value of each of the two remaining "lemonade" mixtures and record your prediction for each in the appropriate space on the chart.

8. Using the procedure described in step 6, determine separately the pH of each of the mixtures in jars #2 and #3. Then record these measured values in the appropriate spaces on the chart.

9. Compare the predicted with the actual pH values of lemon juice/water mixtures in various concentrations. Then, discuss some of the important factors relating to the group activity:

   • which pH value identifies the mixture that has the greatest acidity,
   • which "lemonade" sample would be expected to taste the most tart,
   • how the preliminary determination of pH values for undiluted lemon juice and for plain water proved to be helpful in making predictions about relative pH values of the mixtures.

4–9 CHECKING OUT HOW SEEDS TRAVEL

Motivation: In a variety of ways, nature has ensured that not all seeds will fall to the ground directly below the parent plant. Most species have special designs for scattering their seeds to more distant locations, where the young plants that develop will not be faced with problems caused by overcrowded conditions. An activity that focuses on observing the details of structure of a collection of seeds stimulates critical thinking and helps students reach some conclusions concerning the agent that is involved in the dispersal of seeds of each different design in the collection.

Recommended Grade Level: Grades 4–6

Strategies Involved: Hands-on activity
Student involvement
Group interaction
Science process skills development

Materials Required:

Each group of four students will need:

- a small tray with a collection of assorted seeds (maple, lotus, cocklebur, dandelion, milkweed, blackberry, burdock, violet, beggar-ticks, or others, as available)
- a copy of the SEED STRUCTURES AND THEIR METHODS OF DIS-PERSAL RECORD SHEET

Procedure:

Instruct students, working in groups of four, to proceed as follows:

1. Examine the seeds in the collection and identify those with which some students in the group are familiar.
2. Compare the structure of the seeds, noting instances in which two or more designs show a similar construction.
3. Discuss with members of the group any seed structures that appear to be unique.
4. Look for clues that suggest specific methods by which seeds can travel from the parent plant to a different location.
5. Record in appropriate spaces on the chart all agreed-upon inferences concerning the method of dispersal for which specific seeds are suited.
6. Describe the structure that assists in the dispersal of three additional kinds of seeds and add the names and method of dispersal of these seeds to the chart.
7. Then, write a group statement that explains why seeds have an advantage if they are specially constructed to be dispersed by agents such as wind, water, animals, mechanical devices, and humans.

Name _____   Date _____

SEED STRUCTURES AND THEIR
METHODS OF DISPERSAL
RECORD SHEET

Name of Seed	Description of Seed Structure	Method of Seed Dispersal
Maple	winglike structure	wind
Dandelion	small parachute-like tufts of fibers	
Milkweed		
Beggar-ticks		

4–10   CHANGING THE VOLUME OF A CUPFUL OF APPLES

Motivation: When does a cup full of food not contain a cupful of food? Students can be stimulated to develop critical thinking skills by inferring and predicting what possible changes in volume will occur when a cupful of apple slices is crushed to produce an applesauce-like consistency. Their predictions, backed by scientific support, can then be checked for accuracy when the apple slices are spoon-crushed and remeasured in the cup.

Recommended Grade Level: Grades 4–6

Strategies Involved: Hands-on activity
Student involvement
Group interaction
Science process skills development

Materials Required:

Each group of four students will need:

- one cupful of apple slices from soft apples, peeled and cored
- a strong spoon or potato-masher of suitable size
- a widemouthed measuring cup

Procedure:

Instruct students, working in groups of four, to proceed as follows:

1. Fill a measuring cup with apple slices from soft apples that have been peeled and cored.
2. Examine the cup and its contents and predict any anticipated changes in volume that would be expected to occur if the apple slices were to be crushed.
3. When all group members have reported their predictions, select one member of the group to use a strong spoon to crush the apple slices in the cup.
4. After making certain that no apple material leaves the cup or is left adhering to the spoon, examine the apple material again and measure the contents of the cup.
5. Discuss the accuracy of the predictions made by group members and note any clues (e.g., spaces between apple slices that become filled with smaller particles of crushed apple) that proved to be helpful in making a scientifically based prediction.

6. Suggest methods by which a change in the actual amount of apple material in the cup could be determined. Then plan an activity that would allow for making predictions and determining the weight of apple material in the cup before and after crushing.

4–11   USING SCIENTIFIC EVIDENCE TO SUPPORT A BELIEF

Motivation: Scientific knowledge has progressed along many fronts, resulting in an understanding of many phenomena that were previously explained away by unsupported beliefs. Yet, today some people still believe that "maggots come from garbage" or that "fish extract oxygen for their breathing from the molecules of water in their environment." Debating the issue with a person who holds one of these views stimulates students to engage in critical thinking. In a two-part activity, students can prepare their presentation, based on scientific evidence, and then debate the issue with a science teacher playing the role of devil's advocate. Students should be advised that any of their statements based on information that is scientifically inaccurate and any of their reasoning based on logic that is scientifically faulty will be challenged.

Recommended Grade Level: Grades 7–8

Strategies Involved: Student involvement
                         Student interaction
                         Science process skills development

Materials Required:

- student laboratory notebooks
- textbooks and classroom and library reference materials

Procedure:

Involve all students in the following:

Part 1:
  1. After students have engaged in an activity focusing on breathing methods used by animals that live in water, present them with a challenging situation:

     "A teacher, whom they all know, has stated that fish get oxygen for their breathing by breaking down molecules of water (H_2O) into hydrogen and oxygen."

  2. Allow students to review science textbook and notebook material related to this topic.
  3. Then permit students to meet in small groups to research the topic further and to plan a strategy for presenting convincing arguments in support of a different belief:

     "Molecules of oxygen that are dissolved in the water supply fish with oxygen needed for their breathing."

Part 2:
  1. Enlist the aid of a science teacher from a different classroom to play the devil's advocate in a student/visiting teacher debate on the topic.
  2. List on the chalkboard all points made by the teacher and by the students for a later comparison of the merit of statements made in support of their respective positions taken in the debate.

3. Involve all students in a discussion that focuses on scientific evidence offered in support of the statements made by the teacher and by the student opponents in the debate.

   (Ex.) *Teacher statements:*

   "Fish can live successfully in a balanced aquarium that is not in contact with outside air."

   "Water passes over the gills of a fish during the breathing process."

   Student statements:

   "An aerator is used in the classroom aquarium to bubble air into water in the tank for fish to breathe."

   "Fish come to the surface to gulp air when the dissolved oxygen supply in the water is low."

   "The amount of water does not decrease when fish are present in an enclosed ecosystem."

   "There is no evidence of hydrogen escaping from a body of water when fish are present."

   "Bubbles can be seen rising from water plants placed in an aquarium with fish."

4. Then, allow students to evaluate the two presentations and decide which side presented the more convincing arguments, supported by relevant scientific evidence.

4–12 CHECKING THE USEFULNESS OF A VERSATILE MATERIAL

Motivation: Objects are often wrapped with plastic foam or other materials, such as paper and corrugated cardboard, to prepare them for storage or for shipping and mailing. In addition to acting as a protective cover, which reduces the risk of breakage, shock, or upset, these materials help to protect delicate objects and substances from damage due to temperature changes in very hot or very cold weather.

Recommended Grade Level: Grades 6–8

Strategies Involved: Hands-on activity
Student involvement
Group interaction
Science process skills development

Materials Required:

Each group of four students will need:

- two soup cans, cleaned out and with tops removed
- two thermometers
- rubber bands

- hot water
- a rectangular sheet of plastic foam from packaging
- a Magic Marker
- a copy of the chart TEMPERATURE OF HOT WATER STORED IN CANS WITH AND WITHOUT INSULATION

Procedure:

Instruct students, working in groups of four, to proceed as follows:

1. Obtain two clean soup cans, with tops removed and no jagged edges remaining.
2. Cover the side walls of one can by wrapping a layer of plastic foam around the outside of the can and holding it in place with the use of rubber bands.
3. With a Magic Marker, label the wrapped and unwrapped cans #1 and #2, respectively.
4. Prepare both cans, #1 and #2, as follows:

   - Fill each can with hot water from the hot water faucet.
   - Position both cans conveniently on a work area table.
   - Insert a thermometer in each can.

5. Take temperature readings of the water in each can and record the temperature readings on a chart that provides separate columns for cans #1 and #2.

TEMPERATURE OF HOT WATER STORED IN CANS WITH AND WITHOUT INSULATION		
Time	**Temperature in Can #1 with Insulation**	**Temperature in Can #2 Without Insulation**
Starting time		
After 2 minutes		
After 4 minutes		
After 6 minutes		
After 8 minutes		
After 10 minutes		

6. Record five additional temperature readings for each can, spacing readings at 2-minute intervals.
7. Compare the temperature readings recorded for the water in the two cans during the same time period.
8. Suggest reasons for differences noted in the rate of cooling of water in the two cans, one with and one without a plastic foam wrapping.
9. Prepare a group statement concerning the effectiveness of plastic foam, as illustrated in the activity, and list six specific cases in which it would prove to be useful.

5

Developing Scientific
and Technological
Literacy and Competency

There is no question but that human lifestyles have changed over the years, and that these changes have come about primarily as a result of our own curiosity and ingenuity. With each generation, new scientific knowledge has been gained and put to use in the form of technologies to solve current problems and to make life easier and more enjoyable.

Science and technology have always been interacting forces in our search for solutions to the problems faced by our society. After having discovered fire, for example, primitive people decided that it was to their advantage to enjoy the benefits and comforts it provided in caves, despite the accompanying smoke that filled their homes. In time they devised a way to divert the smoke from the inside of the cave, solving the problem—temporarily—until the ensuing conditions and the ingenuity of the human inquiring mind interacted, resulting in the discovery of better sources for producing heat.

Down through the ages the technique for problem solving has followed the same basic pattern—

> . . . the employment of a device or method growing out of a new discovery is implemented, but is found to be accompanied by an undesirable feature or side effect . . . the curious and inquiring human mind examines the situation to discover why the problem exists and to explore possible ways to overcome it . . . if the advantages outweigh the disadvantages and there is a feasible way to overcome the problem, the correction is made and people live with their

discovery until conditions and their ingenuity and resources interact to find a better way, enabling them to replace the original discovery . . .

—thus coming full circle in a repeating pattern.

Although only a small percentage of the present school population is expected to enter the science/technology sector, it is important that all citizens be informed about basic scientific principles and their use in technological processes and services that affect our lives and the environment. It is also important that students increase their scientific and technological literacy, commensurate with their ability to comprehend the scientific principles involved in relevant technologies, and that they be given opportunities to make informed decisions about the "options." Science programs at all levels need to recognize this as a primary goal.

ACTIVITIES FOR DEVELOPING SCIENTIFIC LITERACY

5-1 ANALYZING THE ACTION OF A FIRE EXTINGUISHER

Motivation: Although water, sand, or salt may sometimes be used for fighting a fire, most homes and public places rely on the use of chemical fire extinguishers for putting out small fires. Among them, the carbon dioxide extinguisher is the most popular, primarily because of the convenience with which it can be handled and the effectiveness with which it controls a variety of different kinds of fires without causing further damage to property. It consists of a strong steel cylinder that contains liquid carbon dioxide maintained under pressure at a temperature of -80 degrees C., so that when the valve is opened, the liquid rushing through the cone-shaped nozzle expands rapidly, changing its form to a white frost, which quickly covers the burning material. The resulting twofold action, which both cools and blankets the fire, effectively extinguishes the blaze. The smothering action of carbon dioxide can be investigated when heavy carbon dioxide gas settles around the flame of a lighted candle, shutting off its oxygen supply and quickly extinguishing the flame.

Recommended Grade Level: Grades 4–6

Strategies Involved: Teacher demonstration
Student/teacher interaction
Science skills development

Materials Required:

- a candle
- a jar top
- a bottle
- a measuring spoon
- baking soda
- white vinegar
- paper
- matches

Procedure:

Conduct an activity in which all students can participate:

1. On a table top that permits clear viewing by all students, perform the following:

   - Light a candle and allow a few drops of melted wax to fall onto an inverted jar top.
   - Set the base of the candle on the melted wax and hold it securely until the candle becomes firmly anchored.
   - Place ½ tablespoon of baking soda in a bottle.

- Add 2 tablespoons of white vinegar to the baking soda in the bottle.
- Roll a piece of paper to form a trough and direct its lower end toward the candle flame.
- Then, tilt the bottle and hold it at a slant along the other end of the paper trough.

CAUTION: DO NOT ALLOW ANY OF THE BAKING SODA/VINEGAR MIXTURE TO FLOW OUT OF THE BOTTLE.

2. Relight the candle and repeat the procedure to enable all students to make close observations of the activity. Then, engage all students in a class discussion in which they report their observations and respond to related questions:

- What happened when vinegar was mixed with baking soda?
- What happened to the candle flame when the mouth of the bottle was tilted over the flame?
- Was the gas produced by the chemical reaction lighter or heavier than air?

3. Extend the discussion by encouraging students to identify:

- an important characteristic of the carbon dioxide gas that was produced by the chemical reaction that occurred,
- reasons why the CO_2 (carbon dioxide) fire extinguisher is recommended for extinguishing fires involving burning grease, oil, and gasoline, but not for large-scale forest fires.

5-2   TRACING THE DEVELOPMENT OF A TASTY TREAT

Motivation: The popularity of popcorn as a snack food among young people on visits to amusement parks, or while watching television, gives the impression that a new and novel product has recently been developed. Actually, it is one of the oldest forms of corn. It was grown and used by the American Indians many centuries before the arrival of early explorers from European countries. Today we know that its unique characteristic of popping when heated is due primarily to the moisture content within the seed. Normally at a level of 19 percent when the corn is harvested, the moisture content of the kernels is reduced by a slow drying process to produce a perfect popping moisture content of 13½ percent. When the corn kernels are heated, this water is then converted to steam, which builds pressure within the hard coat of the kernel and causes it to burst and expand, forming a ball of white fluff, which is enjoyed as a delicious snack treat.

Recommended Grade Level: Grades 4–8

Strategies Involved: Hands-on activity
Student involvement
Student interaction
Science skills development

Materials Required:

Each group of eight students, forming two teams (A and B) consisting of four students each, will need:

- 40 kernels of popping corn (20 kernels per team)
- two hot plates (one per team)
- two pyrex beakers with stoppers (one per team)
- two hot hands* or other hand-protectors (one per team)
- two plastic bags (one per team)
- several pins (team B only)
- cooking oil

Procedure:

Instruct each group of eight students to form two teams (A and B) and to proceed as follows:

1. Begin the activity:

   - Team A, obtain 20 undamaged kernels of popping corn.
   - Team B, obtain 20 kernels of popping corn and penetrate the seed coat of each by piercing with a pin.

2. Proceed with the activity:

   - Team A, using the undamaged corn kernels:

   1) Place cooking oil in the beaker so that the bottom of the beaker is coated with a very thin film.
   2) Add 20 kernels of popping corn to the beaker.
   3) Place the beaker on a hot plate that is set at a medium temperature setting.
   4) Watch the beaker and observe what happens to its contents.
   5) Using a "hot hand," or other hand-protector for handling hot glassware, remove the beaker from the hot plate.

CAUTION: DO NOT TOUCH THE HOT BEAKER WITH YOUR BARE HANDS!

   6) Transfer the popped corn kernels to a plastic bag.

*A Scienceware product, available from Bel-Art Products, Pequannock, New Jersey 07440.

- Team B, using the corn kernels assigned, follow the procedure as outlined in #2, steps 1–6, for Team A.

3. Compare the results obtained by teams A and B:

- What effect did heating have on the corn kernels?
- How did the results obtained with pierced seed coats compare with those obtained with kernels that had not been pierced?
- Why is pressure necessary for "popping?"

4. Based on the findings in this activity, write a short paragraph explaining an essential difference between an improved form of popcorn that expands 40 times when popped over kinds that expand only 25 times.

5–3 UNDERSTANDING "WINDCHILL FACTOR"

Motivation: On a cold, blustery day it is not unusual for a person to feel that the temperature is lower than that indicated by the thermometer. This is because the faster the wind blows, the faster the body loses heat. Using temperature alone as a basis for the selection of clothing appropriate for children engaged in outdoor winter activities is not only unwise; at times, it can be dangerous.

During the winter months, therefore, weather reports usually include a statistic known as the windchill factor, which is the temperature in still air that would have the same cooling effect as a given combination of temperature and wind speed. For example, when the wind is blowing at a rate of 20 miles per hour, an actual temperature reading of 0° F. has a windchill factor of −40° F. This windchill temperature is not an exact index of how cold it feels to an individual, since other factors besides temperature and wind speed are, of course, also involved. However, it does give a better idea of how cold an individual feels than does the temperature alone. The U.S. National Weather Service has prepared tables for converting a given combination of temperature and wind speed to its equivalent still-air temperature, namely the windchill factor.

Recommended Grade Level: Grades 6–8

Strategies Involved: Hands-on activity
Student involvement
Student interaction
Interdisciplinary approach
Science skills development

Materials Required:

Each student/partner combination will need:

- a sheet of construction paper
- a copy of an EQUIVALENT WINDCHILL TEMPERATURES chart

Procedure:

Instruct students, working with partners, to proceed as follows:

1. Fold a sheet of construction paper to form an accordion fan.

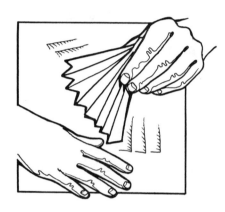

2. In a room where the air is relatively still, extend one hand in front of you and allow your partner to wave the paper fan a few centimeters above your hand.
3. Note the sensation created by the air as it moves above your hand.
4. Ask your partner to increase the speed of the moving air by waving the fan faster and more briskly. Note any differences in the sensation experienced.
5. Then, in order that your partner can also experience firsthand the sensations on the skin caused by air moving at different speeds, change places with him/her and repeat steps 2–4.
6. Compare the sensations you experienced with those of your partner and relate the experiences to:

   - the fact that one feels much colder on a chilly, windy day when the wind causes heat to escape from the body more rapidly than on a day when the temperature is the same but the air is relatively calm,
   - the value of winter weather reports that provide an estimate of how cold the combined temperature and wind speed make a person feel:

     1) Using the EQUIVALENT WINDCHILL TEMPERATURES chart below, tell how cold it feels when the wind is blowing at 10 miles per hour and the reading on the thermometer is 20° F.
     2) Tell how the speed at which the wind is blowing affects how cold a person feels at a specific temperature reading as shown on a thermometer.

EQUIVALENT WINDCHILL TEMPERATURES							
Wind (mph)	**Thermometer Reading in Degrees F.**						
calm	**30**	**20**	**10**	**0**	**−10**	**−20**	**−30**
5	27	16	7	−6	−15	−26	−35
10	16	2	−9	−22	−31	−45	−58
20	3	−9	−24	−40	−52	−68	−81

7. Based on your investigation, write a short explanation telling:

- why the speed of the wind has an effect on how cold a person feels at any given temperature,
- why a cool breeze on a hot day makes one feel more comfortable and "refreshed,"
- why the windchill factor is more appropriate for use in weather reports during the winter months than during the summer,
- why the windchill estimates are useful as a guide for safeguarding one's body against overexposure to the cold.

5–4 EXAMINING A WHIRLPOOL PATTERN

Motivation: The effects of the earth's rotation from west to east are many and widely varied. For example, it causes changes in direction of air and water currents, both north and south of the equator, and exerts an identifiable influence on the path of water rushing down a bathtub drain. In the northern hemisphere water usually drains in a counterclockwise direction, while in the southern hemisphere the direction is usually clockwise. Many factors in the immediate environment, however, are responsible for causing some inconsistencies and reversals in the expected pattern.

Recommended Grade Level: Grades 4–8

Strategies Involved: Student involvement
Student/teacher interaction
Science skills development

Materials Required:

- a classroom sink
- a large quantity of water to be discarded
- a chalkboard
- classroom and library reference materials

Procedure:

Conduct a class activity that involves all students:

1. When the aquarium water is due to be changed, pour a bucket of the discarded water into a sink while students observe what happens.
2. Encourage students to comment on the pattern observed as water rushes down the drain.
3. Record information on the chalkboard.
4. Repeat the procedure several times, making note of variations and/or consistencies in the pattern identified.

5. Ask students to make an at-home observation—dishwater draining from the sink or bathwater from the bathtub—for identifying the direction of a whirlpool pattern in a specific drainage situation.

6. On the next day, ask students to report their at-home findings and list the patterns observed for each case on a chart on the chalkboard.

7. Discuss all class findings:

   • Compare the number of cases reporting a counterclockwise whirlpool with those reporting no evidence of a counterclockwise motion.

   • Consider possible reasons for variations reported:

     1) temperature of the room,
     2) air currents in the room,
     3) other motion of the water being observed.

   • Incorporate information from reference materials.

8. Pose a question that asks students to make a prediction based on their understanding of the activity:

   "A man who travels widely has observed that water rushing down a drain in the northern hemisphere normally swirls in a counterclockwise direction, while in the southern hemisphere the direction is usually clockwise. What would you tell him to expect about the direction of water being drained from a bathtub if he were to drain the tub on shipboard at the same time the ship was crossing the equator?

   What explanation could you offer if his observations were not consistent with your prediction?"

5-5 DETERMINING THE EFFECT OF PRESSURE EXERTED ON SNOW

Motivation: In some regions of the world great masses of snow and ice cover the earth's surface. In these areas, where more snow falls each year than melts, the snow accumulates, often forming thicknesses of up to 1 mile. Then, in much the same way that a snowball, when pressed between a person's hands, turns to a ball of ice, the weight of the upper layers of snow exerts pressure on the lower layers to cause the formation of a heavily compacted layer of ice at the bottom. It is this ice layer beneath the snow that becomes a river of ice, known as a glacier.

Recommended Grade Level: Grades 4–6

Strategies Involved: Hands-on activity
Student involvement
Science skills development

Materials Required:

Each group of four students will need:

• a cleaned-out olive jar
• five or six marshmallows

- a cardboard disc of slightly smaller diameter than that of the jar
- several small, heavy weights

Procedure:

Instruct students, working in groups of four, to proceed as follows:

1. Place five or six marshmallows inside an olive jar to form a vertical stack.
2. Place a cardboard disc on top of the stack of marshmallows.
3. Place a weight on top of the disc and observe what happens.
4. Add additional weights and observe the effect that is produced.
5. Discuss with other members of your group what happened to the marshmallows when pressure was applied from above:

   - Is there evidence that the marshmallows spread outward?
   - Is there evidence that the marshmallows stick together and solidify when they are compressed?
   - What would have happened if the sides of the jar had not been there to hold the marshmallows within the jar?
   - If the marshmallows had been snow, what form would they have assumed as a result of the pressure applied from above?

6. Then, recall a snowball fight in which you engaged and describe how you converted a snowball to an iceball.
7. Write a paragraph explaining how this activity illustrates the process by which glaciers are formed.

5–6   LOCATING THE RELATIVE POSITIONS OF OBJECTS ABOVE THE EARTH'S SURFACE

Motivation: Distances from one location on earth to another are easily comprehended in terms of a short distance traveled by bike between home and school, or in terms of a long distance involved in the postal delivery of a letter from a pen pal who lives halfway around the world. It is more difficult to visualize the distance from the earth to the sun or the moon, or to some extraterrestrial event occurring that involves either natural phenomena or man's efforts in space exploration.

By compressing a great distance to a small linear scale it is possible to locate the relative positions of some objects in space and to mark the distance at which some events occur above the earth's surface. A visual representation of distances in space, reduced to scale and taped to the classroom floor, enables students to develop a concept of the enormous distances that separate Earth from other locations in the universe.

Recommended Grade Level: Grades 7–8

Strategies Involved: Hands-on activity
Student involvement
Science skills development
Interdisciplinary approach

Materials Required:

Each group of four students will need:

- one 10-foot length of adding machine tape
- masking tape cut into several 2-inch strips
- a 12-inch ruler
- colored marking pens
- several balls of varying degrees of smallness
- a copy of the chart "Approximate Distances of Objects Above the Earth's Surface," and other suitable reference material.

Procedure:

Instruct students, working in groups of four, to proceed as follows:

1. Clear a floor area that is more than 10 feet long.
2. Roll out the strip of adding machine tape and apply 2-inch strips of masking tape at intervals along its length to secure it to the floor.

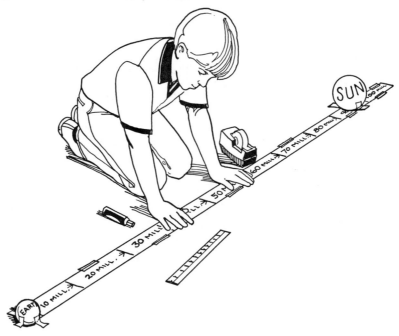

3. Place a small ball at one end of the tape to represent the location of the planet Earth.
4. Using a ruler, mark off ten 1-foot segments along the paper strip.
5. Allowing the entire strip to represent a distance of 100,000,000 miles, label the distance of each segment to show its distance from Earth.
6. Locate the position of the sun, as represented on the paper strip and, using a pen of contrasting color, indicate this distance in miles. Then place a ball, larger than "Earth," at this location.
7. Similarly, locate the position of other objects shown on the distance chart, and place balls or other markers accordingly.
8. Examine your distance chart and discuss it with other members of your group:

   • What would you estimate to be the position of Alpha Centauri, the nearest star to Earth after the sun?
   • Why are distances from Earth to stars beyond the sun usually measured in "light-years?" (A light year is the distance traveled by light in one year.)
   • Estimate the distance traveled on a trip to the moon in terms of the number of trips around the earth at the equator. (The distance around the earth at the equator is 25,000 miles.)

APPROXIMATE DISTANCES OF OBJECTS ABOVE THE EARTH'S SURFACE	
Object	**Distance**
Sun	93,000,000 miles
Moon	240,000 miles
Mercury	50,000,000 miles (minimum distance)
Venus	25,000,000 miles (minimum distance)
Mars	35,000,000 miles (minimum distance)
a man-made satellite	300 miles
northern lights	100 miles
top of Empire State building	1,250 feet

9. Write a group report in which you explain the advantage of a scale model for studying something that is extremely large or extremely small.

ACTIVITIES FOR DEVELOPING TECHNOLOGICAL LITERACY

5-7 EVALUATING THE EFFECTIVENESS OF ULTRAVIOLET LIGHT AS A FOOD PRESERVATIVE

Motivation: Like other forms of life, microorganisms are sensitive to many factors in their physical environment. Although they are widely distributed on earth, each species of microscopic life is limited to a relatively narrow range of light, temperature, moisture, chemical pH, and atmospheric conditions that it finds to be favorable.

Man can derive much benefit from the judicious application of this knowledge. For example, by exposing some bacteria, fungi, and mold spores to ultraviolet light, using relatively low levels of intensity and short exposure times, he can effect a 20 to 80 percent kill. Such efforts contribute to both the control of disease and the reduction of food spoilage caused by microorganisms.

Recommended Grade Level: Grades 7–8

Strategies Involved: Hands-on activity
Student interaction
Science skills development

Materials Required:

Each group of four students will need:

- two plastic petri dishes
- filter paper to fit the inside base of the petri dishes
- bread mold spores collected from moldy bread
- one-half slice of fresh bread, with no preservatives added
- an ultraviolet light source
- safety goggles for each student
- water
- classroom and library reference materials
- teacher access to school health records

NOTE: Identification of students with allergies and asthma should be made by consulting school health records. These students can be official record-keepers for the observations made in this activity. Do not let them breathe any mold spores.

- a hand lens or magnifier (optional)

Procedure:

Instruct students, working in groups of four, to proceed as follows:

1. Prepare two petri dishes:

   - Place moistened filter paper in the base of each petri dish.
   - Cut one-half slice of bread in half.

- Place one bread piece on the moistened filter paper in each petri dish.
- Sprinkle water over the bread to moisten its surface.
- Sprinkle mold spores over the surface of the moistened bread.

CAUTION: WEAR SAFETY GOGGLES WHILE PERFORMING THE NEXT STEP IN THE PROCEDURE.

- Expose one prepared petri dish (uncovered) to ultraviolet light for a period of 10 minutes.

2. Remove the safety goggles and continue:

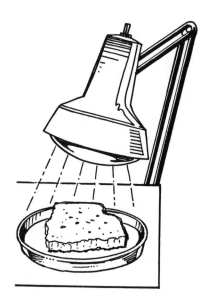

- Place the covers on the petri dishes and label each appropriately: "exposed" or "nonexposed."
- Place both petri dishes in a warm, darkened location where they will not be disturbed.
- Examine the contents of each petri dish on a daily basis and watch for evidence of mold growth on the bread surface.
- Record observations made over a 10-day period.

3. Discuss with members of your group:

- differences noted between the two samples, with possible reasons for the presence or absence of mold growth,
- reasons for using a sample that has not been exposed to ultraviolet light, as well as the light-exposed sample of spores planted on moistened bread,
- reasons for wearing safety goggles during the "exposure" interval,
- effectiveness of ultraviolet light in retarding food spoilage,
- reasons for using bread that has not been treated with any other preservative,
- the advantages of ultraviolet light treatment of bread over the use of chemical additives to retard the growth of mold.

4. Based on the discussion of the activity and the use of reference sources for related information, prepare a list of advantages and disadvantages associated with the use of ultraviolet light to retard the spoilage of foodstuffs.

5–8   CHANGING ONE FORM OF ENERGY TO ANOTHER

Motivation: Like matter, energy changes from one form to another. We see examples of light energy changing to chemical energy in photography and in the fading of the bright colors of a banner. Light energy changes to heat energy when solar heaters are used to heat homes, to electrical energy when photoelectric cells issue warning signals that intruders have entered restricted areas, and to mechanical energy when the rotating blades of a radiometer are placed in direct light. Technology also enables us to program more than one energy transformation before the desired form is employed to accomplish the designated work. Many such series of transformations, starting with a primary source, can be investigated.

Recommended Grade Level: Grades 4–6

Strategies Involved: Hands-on activity
Student involvement
Student/student and student/teacher interaction
Science skills development

Materials Required:

- an assortment of devices selected and/or brought in by students, such as a battery-operated toy car or truck, a flashlight, a record or tape player, an electric hot plate with an automatic stirrer, and other devices
- copies of the TRANSFORMATION OF ENERGY RECORD SHEET (one per student)

Procedure:

Involve all students in the activity:

1. Encourage students to demonstrate their devices to other members of the class.
2. Permit time for students to record observations and make determinations of energy transformations involved in each demonstration.
3. Discuss the demonstrations and observations made:

- Compare the information the students have recorded on their charts.
- For each device demonstrated, ask the student to:

  1) present evidence of the primary energy source and its transformation into a usable energy form,
  2) give evidence of intermediate forms of energy, if any,
  3) give evidence of secondary energy, if any.

- List some applications of the Law of Conservation of Energy to some of the products of man's technology.

TRANSFORMATION OF ENERGY RECORD SHEET

Name of Device	Kind of Energy in Stored Form	Primary Kind of Energy After Transformation	Secondary Kind of Energy (if any)
1. Flashlight	Chemical	Light	Heat
2. Walkman			
3. Toy truck			
4.			
5.			
6.			
7.			
8.			
9.			
10.			
11.			
12.			
13.			
14.			
15.			
16.			
17.			
18.			
19.			
20.			
21.			
22.			

5-9 INCREASING THE POWER OF A MAGNET

Motivation: Throughout the ages man has used available technologies to make his work easier. For example, there is much evidence to indicate that the wheel and axle, the inclined plane, and other simple machines contributed significantly in the building of ancient bridges, roadways, and the pyramids.

Today this is still true; a familiar and very practical example is the electromagnet, which is commonly used in construction work to hoist heavy loads of steel girders and in junkyards to pick up old cars and scrap metal.

Recommended Grade Level: Grades 4–8

Strategies Involved: Hands-on activity
Student involvement
Science skills development

Materials Required:

Each group of four students will need:

- two flashlight "C" batteries
- a heavy iron nail, 10 cm long
- a wire stripper or pliers
- a metric measuring tape
- insulated bell wire (22 or 24 gauge)
- tape
- paper clips
- a metric ruler

Procedure:

Instruct students, working in groups of four, to proceed as follows:

1. Tape two flashlight batteries together, top-to-bottom fashion, so that the metal top of one battery rests firmly against the flat metal bottom of the other.
2. Strip 2.5 cm of insulation from each end of a 4-m length of bell wire.
3. Leaving the first 7.5 cm of wire free, turn the wire in a clockwise direction around a large nail to form a tightly wound coil of 100 or more turns around the shaft of the nail. Leave the last 7.5 cm of wire free also.
4. Attach the coil to the batteries by taping one free end of the wire to the top of the upper battery and the other free end to the bottom of the lower battery. Check to be sure that the metal of the wire makes contact with the metal of the batteries.

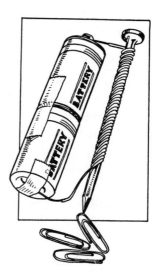

5. Hold the coil with the point of the nail close to a collection of paper clips on the table top and observe what happens.
6. Count the number of paper clips that were picked up.
7. Detach the wire from the top battery and observe what happens.
8. Discuss with members of your group, focusing on:

   - the advantages of an electromagnet over a regular magnet,
   - how an electromagnet can be made stronger or weaker.

9. On the basis of your investigation, make a prediction about the number of paper clips that would be picked up if your coil of wire had only 50 turns and check out the correctness of your prediction.

   Then make a prediction about the number of paper clips that would be picked up if the coil had 150 turns and check out the correctness of that prediction also.

5–10   GIVING BASE METALS AN APPEARANCE OF PRECIOUS METALS

Motivation: A gold medal is not usually made of solid gold, nor is a silver spoon usually pure silver. It has become a common practice to coat some objects made of an inexpensive metal with a thin layer of gold or silver, giving the appearance of the richness and luster of the precious metals, but at a much lower cost. The process involves the use of an electric current and a solution of the metal that is being plated in the form of an incredibly thin layer on the desired object. The small amount of precious metal used in this process accounts for the tendency of the finish on a silver-plated spoon or a gold-plated ring to wear away and reveal the kind of metal that was used in the basic construction of the object.

Recommended Grade Level: Grades 7–8

Strategies Involved: Hands-on activity
                         Student involvement
                         Science skills development

Materials Required:

Each group of four students will need:

- insulated copper wire
- a widemouthed quart jar
- a pair of strong scissors
- two 1.5-volt dry cell batteries
- a small silver-plated spoon that has been discarded
- a small key
- a silver nitrate solution*, prepared by dissolving 85 g of silver nitrate ($AgNO_3$) in 1 liter of warm water
- silver polish and a soft cloth

*Accidental spills of silver nitrate solution should be cleaned up immediately with paper toweling, followed by a thorough rinsing of the affected area with clear water to remove all traces of the chemical.

Procedure:

Instruct students, working in groups of four, to proceed as follows:

1. Use the silver polish and soft cloth to remove all traces of tarnish from the silver-plated spoon.
2. Cut three pieces of insulated wire of appropriate length and remove sufficient insulation from both ends of all three pieces to allow the following connections:

   - Use wire #1 to connect the negative terminal of one dry cell to the positive terminal of the other.
   - Wrap one end of wire #2 around the spoon handle and attach the other end to the unattached positive dry cell terminal.
   - Tie one end of wire #3 through the hole in the key and attach the other end to the remaining unattached negative dry cell terminal.

3. Suspend the key and spoon inside the jar in positions that are opposite one another. If necessary, bend the wires over the rim of the jar to keep the key and spoon well-separated so that they cannot touch each other.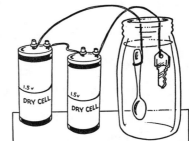
4. Carefully, pour the silver nitrate solution into the jar, covering the spoon and key completely.

   CAUTION: DO NOT SPILL OR SPLASH SILVER NITRATE SOLUTION. IN CASE OF AN ACCIDENTAL SPILL WASH HANDS IMMEDIATELY AND NOTIFY THE TEACHER.

5. Observe what happens to the key.
6. Discuss the activity with other members of your group:

   - Describe the appearance of the "plated" key.
   - Estimate how much of the key is now silver.
   - Determine where the silver came from.
   - Explain why a brass key plated with silver is stronger than a key made of pure silver.
   - Name some other metals that could be used in place of silver to plate an object, such as a key.
   - Explain why this process is called "electroplating."

7. Make a summary statement about the process of electroplating, including:

   - two advantages of the process,
   - three practical applications of the process.

5–11 RECORDING EARTH DISTURBANCES

Motivation: There is much evidence to indicate that the earth's crust is not stable and unvarying, but in a constant state of change and evolution. Some changes occur slowly and are not readily observed. Cracking appears at the surface, upward thrusting results in a slow and gradual buildup of mountains, and general reshaping of rock formations goes on constantly within the earth's crust. Major changes, taking place over millions of years, have formed the topography of the continents and ocean floors. Clearly tremendous forces are at work. At times these forces cause rapid and dramatic changes, as in the case of earthquakes when the earth is made to shake violently. Then the sudden shifting of the earth is accompanied by waves that radiate outward from the point of focus and can be detected at great distances from the actual center of activity. Data collected by sensitive instruments are used to determine the location of greatest severity of the quake.

Recommended Grade Level: Grades 6–8

Strategies Involved: Hands-on activity
Student involvement
Science skills development

Materials Required:

Each group of four students will need:

- two standard masonry bricks
- a ring and ringstand
- masking tape
- sturdy cord or twine
- smooth paper
- a sharpened pencil
- classroom and library reference materials

Procedure:

Instruct students, working in groups of four, to proceed as follows:

1. Place a ringstand, with ring attached, on a table top and, using masking tape, tape the base of the stand to the table so that the ringstand will not move. Then, tape a sheet of smooth paper to the surface of the base.
2. Position a sharpened pencil between two bricks so that, when held lengthwise, the pointed end protrudes beyond the ends of the bricks. Then, tape the bricks together so that the pencil will not shift its position.
3. Use sturdy cord or twine to tie around the bricks and attach them to the ring, allowing the pencil to touch the paper on the surface of the ringstand base.

4. Allow one member of the group to bump against one side of the table, while other members observe what happens. Repeat the procedure, replacing the paper each time, until all students have been involved and all sides of the table have been tested.

5. Discuss the activity with other members of the group:

   • how the strength of the disturbance caused by bumping against the table is represented by the markings on the paper,
   • how it is possible to record the occurrence of a disturbance at a distance from the place where the disturbance actually takes place,
   • how this kind of mechanism is useful for detecting a violent shaking of the earth at a distant location.

6. Research the design of modern seismographs and explain the advantage of a design that uses paper mounted on a revolving drum over one that records disturbances on a stationary piece of paper.

5–12   INVESTIGATING "COLD" LIGHT

Motivation: We generally associate light with heat, since energy transformations from sunlight result in heat, as do those from an electric light bulb. It is light that produces no measurable heat that is unusual. Fluorescent light produced by fireflies is an example of this "cold" light, as is the glow of certain bacteria and fungi. Both are the result of light emissions that draw off the excess energy the organisms cannot use for their metabolic activities.

Chemical light sticks also produce "cold" light. By combining the enclosed chemical compounds, some electrons are caused to move temporarily to a higher energy level, where the energy is first absorbed, and then released in the form of light as the electrons return to their normal energy level. The eerie glow produced by light sticks makes them popular novelties for party fun, but they can also be counted on to provide light in emergency situations where conventional light sources are not available.

Recommended Grade Level: Grades 6–8

Strategies Involved: Hands-on activity
Student involvement
Science skills development

Materials Required:

   • chemical light sticks* for all students
   • a darkened room

*Cyalume™, available from Edmund Scientific Company, Barrington, New Jersey 08007.

Procedure:

Instruct students to proceed as follows:

1. In a darkened room:

   - Hold the slim plastic tube of a chemical
     light stick (Cyalume) with both hands.
   - Bend the tube sharply.
   - Observe what happens.

2. Discuss with other students:

   - what was observed,
   - what colors were produced,
   - what caused the light to be produced,
   - how long the light remained,
   - evidence, if any, of heat production,
   - evidence of energy transformation.

3. Write a summary statement describing a device that produces light without
   using a flame or electricity and explain how this kind of device could be put
   to practical use.

6

Science/Societal Issues

Man's ability to advance technologically has increased at a rate that is faster than his ability to assimilate these advances. There is a gap between our technological progress and social attitudes, which has widened as our concerns about pollution and carcinogens have grown. Not only are these concerns related to certain advances coming out of the scientific community, but various groups and individuals take issue with the requirements that they use safety belts, submit to inoculations and vaccinations against communicable diseases, and use life-support technologies for terminally ill patients.

Although there are many problems associated with scientific and technological advances, science is a benefactor, not an adversary, of society. However, the rapid rate at which advances and breakthroughs are being made, and the almost immediate availability of resulting consumer products and services, have overshadowed the scientific community's attention to all of the possible consequences of their contributions to society. There is always the danger that scientific advances will upset the natural environment or endanger some rare species. In addition, it is possible that other undesirable side effects might be produced, which, with greater foresight and more careful planning, could be prevented. In meeting society's demands, scientists should weigh the benefits of their contributions against their potential for inflicting harm. When the scientific community does not face up to every aspect of its social responsibility, the credibility of science becomes damaged and the value placed on its contributions is diminished.

The technology that has enabled man to attain his present level of advanced civilization should be considered to be neither totally good nor totally bad; the good and/or bad associated with it can be evaluated only in relation to the conditions surrounding its employment. In response to the growing awareness that man must be more sensitive to the fragility of the balance between the earth and life upon it, the primary goal of technology—to supply better goods and services at

91

low cost—has been broadened to include the conservation of all resources and the protection of the environment. This is a challenge that must be met by the same human ingenuity that continues to produce amazing technological advances at an ever-increasing rate.

Assessing the impact of science on society is a value-laden issue, with implications for various social and governmental agencies as well as for the scientific community. But mostly it is the responsibility of individuals to become informed, to act responsibly, and, as with the making of all value judgments, to recognize the necessity of making trade-offs.

Science/Societal Issues in the Classroom

Students are often enthusiastic about expressing their individual views regarding science-related societal issues. Their interests, of course, change over time. The early middle school student concerns, usually related to baby seals and endangered species, evolve into a growing awareness of broader societal issues, such as genetic engineering, cloning practices, and "right-to-die" questions. Dealing with their current interests and concerns at each level has long-range benefits; it helps students develop a familiarity with the methods used to reach intelligent assessments and decisions concerning societal issues about which they may, one day, be asked to vote.

It must be remembered that value judgments are involved in all assessment and decision-making situations. Therefore, care should be taken to present topics objectively and impartially and to respect all valid conclusions by accepting any decision based on supporting evidence and the use of valid thought processes and reasoning technique—even though it may be unpopular or different from that of the teacher or of other members of the class.

In the activities that follow, students should be guided in their investigation by a procedure that consists of four basic steps arranged in logical sequence:

1. Identify a significant science/societal problem or issue—either on a student-initiated or teacher-motivated basis.
2. Gather pertinent background information and supporting data.
3. Make decisions based on the collected information and data.
4. Plan a strategy for taking action to alleviate the problem.

ACTIVITIES THAT INVOLVE SCIENCE/SOCIETAL ISSUES

6–1 INVESTIGATING A RENEWABLE ENERGY SOURCE

Motivation: It would be difficult to think of what our lives would be like in a world without the comforts and conveniences we have come to enjoy in our daily activities. We take for granted the ease with which we engage in modern modes of communication and transportation, the operation of our industrial and business interests, and the maintenance of our homes with the latest innovations in heating systems, air conditioners, television sets, and cooking appliances. Currently these, and countless other devices used in our homes and outside activities, are powered primarily by energy from fossil fuels, whose supplies on earth may soon be depleted. To avert the dire consequences of an unsolved energy crisis, alternative energy sources must be investigated now, and decisions made only after giving careful consideration to the costs, benefits, and possible risks that are involved in each.

Recommended Grade Level: Grades 4–8

Strategies Involved: Hands-on activity
Group interaction
Science skills development
Critical thinking/problem-solving approach

Materials Required:

Each group of four students will need:

- two identical glass or plastic bottles
- two one-hole rubber stoppers to fit the bottles
- two thermometers
- water
- a marking pen or pencil
- a copy of the chart EFFECT OF DIRECT VS. INDIRECT SUNLIGHT ON TEMPERATURE OF WATER
- a location that provides direct sunlight

Procedure:

Instruct students, working in groups of four, to proceed as follows:

1. On a bright, sunshiny day, hold one hand where it will receive direct sunlight, while the other hand remains unexposed to the direct rays of the sun.
2. Compare the difference in warmth felt by the two hands and think of the many devices requiring heat energy that might operate on heat from the sun.
3. Investigate the ability of a material or object other than your hand to be heated by energy from the sun:

   - Distinguish between two identical bottles by marking them with numbers, #1 and #2.

93

- Fill both bottles with tap water.
- Carefully insert a thermometer through the hole in each of the stoppers.

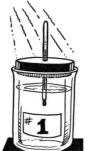

- Place a thermometer in each of the bottles, making adjustments, as needed, to ensure that the thermometer bulbs are at the same level in each bottle and that the stoppers fit tightly at the bottle openings.
- Place bottle #1 in an area that receives direct sunlight, and place bottle #2 in a different location in the room, away from direct sunlight.
- Record starting and 15-minute interval temperature readings for both bottles of water, as provided for on a data chart for use with this activity.

4. Analyze the data chart and discuss key points with other members of the group:

- reasons for the differences noted in temperature,
- how bottle #1 illustrates a basic plan for solar heating in homes,
- advantages and disadvantages of using a renewable energy source, such as solar energy, as compared with those associated with nonrenewable sources, such as fossil fuels.

5. Write a summary statement in which you describe what you have learned about solar energy and tell how you would encourage people to convert from fossil fuels to solar energy for heating their homes.

Name _____   Date _____

EFFECT OF DIRECT VS. INDIRECT SUNLIGHT
ON TEMPERATURE OF WATER

	Temperature in degrees Celsius Bottle #1	Temperature in degrees Celsius Bottle #2
Starting temperature	_____	_____
After placing Bottle #1 in direct sunlight and keeping Bottle #2 in the location away from direct sunlight		
15-minute reading	_____	_____
30-minute reading	_____	_____
45-minute reading	_____	_____
After removal of Bottle #1 from area of direct sunlight and keeping Bottle #2 in its location away from direct sunlight		
15-minute reading	_____	_____
30-minute reading	_____	_____
45-minute reading	_____	_____
60-minute reading	_____	_____

6–2 EXPLORING RECYCLING PRACTICES AS AN ANSWER TO THE SOLID WASTE PROBLEM

Motivation: Our "throw-away" society has been responsible for the creation of many problems. The amount of solid waste produced has been increasing at an alarming rate, the costs for collecting and transporting this waste to available landfills are rising rapidly, and the landfill sites are filling up, with few prospects for further accommodations. The prediction is that most states will be experiencing a major waste disposal crisis before the year 2000. Developing a solid waste management plan that includes the reprocessing of recyclable materials, such as paper, glass, metals, and rubber, will help to alleviate the problems caused by our wasteful practices, which have also proven to be damaging to the environment.

Recommended Grade Level: Grades 4–8

Strategies Involved: Student involvement
Investigative/problem-solving approach
Critical thinking skills development

Materials Required:

- a collection of recyclable materials that have been discarded, such as glass bottles, paper products, plastic bottles, aluminum cans, and newspapers
- classroom and library reference books and materials
- a large, class-prepared chart for the bulletin board

Procedure:

After students have assembled a suitable variety of recyclable materials, instruct them to proceed as follows:

1. Examine the collection of materials that have been assembled and determine how many are represented in the weekly trash from individual student households and from daily trash collections at school.
2. Participate in a brainstorming session to consider:
   - what effect on landfill areas will result from placing these types of articles in containers for a general trash/garbage pickup,
   - what effect on natural resources will result from burying these types of articles with the garbage in landfills,
   - what benefits would be expected to accompany the practice of separating recyclable materials for collection and processing.
3. Make a collection of envelopes, brochures, greeting cards, and other paper products that have been made from recycled paper.
4. Research natural cycles (hydrological, carbon, oxygen, and nitrogen) and prepare a class chart showing how man can recycle materials by following nature's pattern for getting new products from old.
5. Plan a recycling program that offers recommendations for:
   - the kinds of solid waste to be recycled,
   - the things that students can do,

- the things that families can do,
- the things that your school can do,
- the things that your community can do, and
- the things that science and industry can do,

so that all will enjoy the benefits of the program.

6–3   TRACKING A MAJOR SOURCE OF A WIDESPREAD AIR POLLUTANT

Motivation: The release of a pollutant into the environment may have some far-reaching effects. Harmful conditions, resulting as a direct consequence of the release, are often accompanied by additional consequences before the full extent of the harm due to the pollutant is known. Tracing the consequences that result from the release of a pollutant helps students to build an understanding of how environmental hazards grow, become widespread, and affect a large segment of society.

Recommended Grade Level: Grades 6–8

Strategies Involved: Student involvement
Group investigation approach
Student/student and student/teacher interaction
Critical thinking skills development

Materials Required:

- a 10-gallon glass aquarium tank
- a glass cover to fit the aquarium tank
- sand
- wood shavings, wood chips, and/or dry leaf litter
- colored paper bubbles
- marking pens
- colored pushpins
- filter paper or white paper toweling
- tape
- long fireplace matches
- a large jar top

Procedure:

Enlist the aid of student volunteers for performing the following:

1. Place an aquarium tank on a table top where it can be viewed clearly.
2. Prepare a layer of sand to a depth of 2 inches in the bottom of the tank.
3. Place wood chips or other combustible material in a large jar top.
4. Place the jar top containing the wood product or leaf litter on the surface of the sand in the tank.

5. Tape a piece of filter paper or paper toweling to the inside surface of the glass cover in a position that is not directly above the combustible material.
6. Lift the glass top slightly at one end and, using a long fireplace match, ignite the combustible material in the tank.
7. Lower the glass cover so that it rests on top of the tank and allow students to observe what happens inside the tank.

Instruct students, working as a class, to proceed as follows:

1. Report all observations made and list them on the chalkboard.
2. Discuss the observations made and relate them to an actual situation that might occur in nature when air becomes smoke-filled and particles are deposited on surfaces.
3. Arrange on the bulletin board a series of paper bubbles showing the widening consequences of what was represented by the investigation. Include bubbles that illustrate direct consequences of the pollution, followed, in some cases, by effects produced at second and third levels as well. Connected by arrows to indicate the sequential order of consequences, the bubbles might be arranged in a pattern such as:

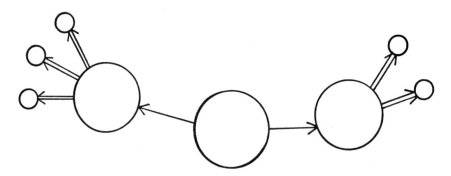

4. Analyze the display of consequences that accompany the form of air pollution investigated and make refinements and adjustments in placement and sequence of bubbles, where necessary.
5. Discuss other aspects of the problem:

   • how air pollution in one area affects other areas,
   • how smoking adds to the problem,
   • how nature adds to the problem,
   • how industry and various modes of travel add to the problem.

6. Consider some possible solutions to the problem of air pollution caused by burning combustible materials:

   • the feasibility of eliminating all burning practices,
   • the use of alternative fuel sources,
   • the use of alternative energy sources.

7. On the chalkboard, make a list of recommendations that would be reasonable to follow and that would not require drastic changes in our lifestyle standards.

6–4 PROBING FOR WAYS TO PROTECT THE OZONE LAYER

Motivation: Recent observations of a deepening hole which occurs seasonally in the upper atmosphere ozone layer over Antarctica have caused scientists to become alarmed. Their concern stems from the fact that any reduction in this protective layer of ozone is accompanied by a similar reduction in the effective filtering out of harmful ultraviolet rays from the sun. The resulting harm to living cells and the increased number of cases of skin cancer among humans are strong reasons for bringing about some changes in the lifestyle of individuals in our society. There is a growing belief that we must find alternatives for the solvents, refrigerants, aerosols, and certain plastic food containers that contain the very chlorofluorocarbon compounds associated with the breakdown of the protective ozone layer.

Recommended Grade Level: Grades 4–8

Strategies Involved: Student involvement
                          Science skills development
                          Investigative approach
                          Making of value judgments

Materials Required:

- an overhead projector and projection screen
- a transparency* showing how the earth's ozone layer filters out some of the ultraviolet light from the sun's rays
- labels from assorted commercial products
- classroom and library reference materials

Procedure:

Project a suitable overhead transparency on the screen and instruct students to proceed as follows:

1. Examine the diagram on the viewing screen and determine:

   - the different kinds of rays that are emitted by the sun,
   - the barrier that prevents some of the harmful ultraviolet rays from reaching the earth,
   - what would be expected to happen if the protective layer were to become even thinner or develop additional holes.

*May be teacher-prepared or purchased from a commercial science materials and supply company.

2. Research information about the ozone layer and the effects of ultraviolet light. Then, participate in a class discussion focusing on:

- harmful effects of ultraviolet rays,
- danger to the ozone layer due to chemical substances called chlorofluorocarbons (CFC), found in aerosol sprays, refrigerants, and some fast-food packaging material.

3. Make a collection of labels from suspect commercial products, such as hairsprays, insecticide sprays, and room fresheners.

4. Read the list of ingredients printed on each label. Then, tabulate on the chalkboard the names of all the chlorofluorocarbon-containing products you investigated.

5. Analyze and discuss your findings:

- Consider the changes that would have to be made in present lifestyles if the use of these convenience products were to be discontinued.
- Consider alternative forms of these products that could be used.
- Consider possible scientific developments that would provide replacements without the loss of convenience associated with CFC-containing products.

6. Based on the evidence, indicating both advantages and disadvantages that accompany the use of CFC-containing products, make a decision about the choice—to continue or discontinue the use of products that contain chlorofluorocarbons.

6–5   BECOMING AWARE OF ENDANGERED SPECIES

Motivation: The list of endangered plants and animals is long and diversified. It includes many species—from the gray whale to the pine barrens tree frog, from the desert poppy in some regions to the large yellow lady's slipper in others. Many endangerments are the result of human exploitation of the natural environment to supply space or food and other consumer products for the ever-increasing world population. Strong opposition to these practices is voiced by individuals and organized groups who would preserve all that is natural in its unaltered form. The question arises: "Can people's needs be supplied by alternative means that would allow all species to survive and pose the threat of extinction to none?"

Recommended Grade Level: Grades 4–8

Strategies Involved: Analysis of data
Group interaction
Critical thinking skills development
Making of value judgments

ENDANGERED SPECIES
AND THE REASONS CITED FOR THEIR ENDANGERMENT

Name of Endangered Species	Reasons Cited for Endangerment
Alaska fur seal	hunted for valuable fur
Blue whale	hunted for meat and oil
Bog turtle	drainage of wetlands
Grizzly bear	overcrowded conditions
Timber wolf	land development
Eskimo curlew	shot for sport
Cheetah	killed for valuable fur
Pine barrens tree frog	drainage of wetlands
European bison	land development
Morelet's crocodile	killed for hides
Asian tiger	killed for sport and fur trade
Texas longhorn cattle	excessive interbreeding
Florida panther	land development
African elephant	killed for illegal ivory trade
Mountain gorilla	destruction of tropical forests
Snow leopard	hunted for valuable pelts
Puerto Rican parrot	habitat destroyed by land development
Dusky seaside sparrow	food chain contaminated by pesticides
Black-footed ferret	spread of deadly distemper infection
Evening primrose	overpicking by wildflower collectors
Wild orchid	market for exotic flowers
Sea turtle	killed for tortoiseshell trade
Desert pupfish	depletion of underground water supply
Everglades kite	drainage of wetlands
Texas blind salamander	interruption of food chain
Mongolian wild horse	interbreeding with domestic horses
Cougar	land development
Yacare caiman	killed for valuable hides
Komodo monitor lizard	competition for remaining food supply
Giant panda	sale of pelts on black market
California condor	killed by hunters
West Indian manatee	slaughtered for meat and hides
Asiatic black bear	overhunting and loss of habitat
Black rhinoceros	poaching for black market horn trade
Atlantic green turtle	captured for meat, oil, and eggs
Asian small-clawed otter	environmental pollution
Coyote	killed by ranchers to protect lambs
Mustang	killed by ranchers to protect livestock
Gray whale	slaughtered for meat and oil
Mexican free-tailed bat	poisonous insecticides entering food chain
Timber rattlesnake	deterioration of natural habitat
Peregrine falcon	infertility due to pesticide poisoning
Blue-spotted salamander	deterioration of natural habitat
Osprey	infertility due to pesticide poisoning
Tuatara	climatic changes

Materials Required:

- copies of the list ENDANGERED SPECIES AND THE REASONS CITED FOR THEIR ENDANGERMENT (one per student)
- reference books and magazines

Procedure:

Distribute copies of the Endangered Species chart and instruct students to proceed as follows:

1. Examine the list of Endangered Species and the Reasons Cited for Their Endangerment.
2. Use reference materials to collect information about additional plants and animals that are endangered and add the names of these organisms to the list.
3. Study the reasons cited for the endangerment of species and discuss:

   - which man can control,
   - which man cannot control,
   - which would be difficult for man to correct without seriously interfering with his lifestyle,
   - what advantages and disadvantages would result from a reversal of some of the situations responsible for the endangerments.

4. Discuss with other members of the class possible ways to resolve the conflict arising out of man's need for more living space, due to his growing population, and the needs of the endangered species.
5. Considering all the evidence, make a statement about man's responsibility toward all endangered species of organisms.

6–6 ANALYZING AN UPSET IN THE BALANCE OF NATURE

Motivation: There are important differences in the kinds of living things that are found in a freshwater pond, a woodland, a desert, or a swampland setting. In each system, the unique combination of life forms is maintained in a delicate balance which can be upset very easily, either by natural means or by human intervention. Such an upset is illustrated by the invasion of African killer bees into the United States, across its southern border, which poses a threat to some beneficial native species. Any introduction of an exotic plant or animal to a different environment may threaten the continued survival of an existing species in a community, or it may place the transplanted species in jeopardy. Because of the risks inherent in the introduction of a new species to an established ecological setting, it is wise to weigh carefully the benefits of such a move against the harm that may result.

Recommended Grade Level: Grades 7–8

Strategies Involved: Data processing skills development
                         Student interaction
                         Identification and organization of satellite issues
                         Making of value judgments based on critical thinking

Materials Required:

Each group of four students will need:

- a large piece of construction or butcher paper
- Magic Markers in assorted colors
- a specimen or picture of one exotic plant or animal
- classroom and library reference materials

Procedure:

Instruct students, working in groups of four, to proceed as follows:

1. Use reference materials to gather information about the plant or animal that your group is investigating.
2. Based on research findings, consider the effects of transplanting the organism from its native habitat to a new and different environment.
3. Prepare a satellite issues poster:

   - Attach a picture of the exotic organism to the bottom portion of a large piece of butcher paper.
   - Above the picture, write the name of the organism within a circle drawn in colored ink.
   - Using a second color, draw lines and well-spaced circles that extend upward from the first circle to represent satellite issues on the second level. Provide one circle for writing each of the immediate consequences of the original transplant.
   - Consider each of the satellites individually and, using a third color, draw lines and circles in which to write the names of satellite issues that arise from each at the second level.
   - Continue this pattern for additional levels, where indicated.

4. Attach the satellite issues poster to the bulletin board and select one group member to explain to the class how the transplantation of the organism was responsible for the snowball effect of many satellite issues at each level.
5. After each group has presented its satellite issues chart for the organism studied by its members, discuss with other students in the class:

   - which transplantations initiated satellite issues that were primarily beneficial,

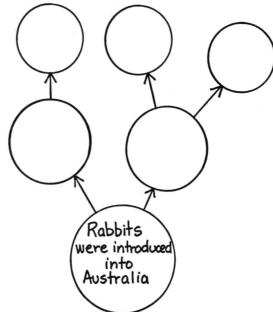

- which transplantations initiated satellite issues that were primarily harmful,
- reasons why transplanting an exotic organism to a new location is more likely to result in harmful than beneficial consequences.

6. Based on the investigation completed and on research involving the organisms in the situations described below, make decisions, supported by scientific evidence, for each situation:

- Should the settlers of Australia have been allowed to import rabbits from their native England?
- Should Mr. Smith be allowed to transfer mallard ducks from North America to New Zealand when he relocates and decides to live in New Zealand?
- Should people be as concerned with the issue of species pollution as they are with pollution of air and water?

6–7 EXAMINING METHODS OF POPULATION CONTROL

Motivation: Although hunting and fishing activities were essential for providing early man with food, modern food-raising methods produce food in such abundance that, today, man need not depend on wildlife forms. Many people oppose the practice of hunting, claiming that animals, too, have a right to live. Others, primarily outdoorsmen who enjoy the sport of hunting and/or fishing, look upon these activities as forms of recreation. The conservationist takes a more objective position, viewing the matter as a scientific problem and examining all aspects of the issue. Considering the effects of man and the environment on the population of wildlife forms gives students a basis for forming a responsible social attitude. Their goal should be to make decisions that would be in the best interest of all populations involved in any issue about which there is much disagreement.

Recommended Grade Level: Grades 4–6

Strategies Involved: Hands-on activity
Student involvement
Student interaction
Decision making based on critical thinking

Materials Required:

- newspaper and magazine articles relating to conflicting views on deer hunting

Also, each group of four students will need:

- two identical planting trays
- garden or potting soil
- a packet of viable seeds
- a favorable location for germinating seeds
- water
- plant markers

Procedure:

Instruct students, working in groups of four, to proceed as follows:

1. Collect information relating to the controversial issue of deer hunting from a variety of sources, such as hunters, members of protest groups, campaign posters, and newspaper and magazine articles.

2. Conduct an investigation with other members of your group—

   • Prepare two identical planting trays:

   1) Place garden soil to a depth of 2 inches in each container.
   2) Moisten the soil to make it workable.

   • Follow the directions on the seed packet for planting the seeds in planter #1 and insert a labeled plant marker in the soil to identify this planting.

   • Sow seeds more generously in planter #2 and label this planting appropriately with a second marker.

   • Place both planters in an area that provides favorable conditions for germination of the seeds and growth of the plants.

   • Set up a watering schedule and continue to provide favorable conditions for both plantings.

   • Observe the growth of plantlets in both planters. Follow directions on the seed packet for thinning the plantlets in planter #1, but allow all plantlets in planter #2 to remain, undisturbed.

   • Discuss with other members of the group:

   1) the appearance and survival rate of plantlets in each of the planters,
   2) the effect of overcrowding on the growth of the plants,
   3) the way in which this investigation relates to the success of a deer herd in an overcrowded population,
   4) the reasons for overcrowded deer populations in some areas,
   5) how hunting (with and without restrictions) would affect the overcrowded conditions and general state of health of the herd,
   6) possible alternatives to hunting as methods for keeping the size of the herd within the capacity of the area to support it.

   • Prepare a list of arguments for and against deer hunting and judge which position, based on supporting evidence, is the more convincing.

6–8 DEBATING THE USE OF ANIMALS IN RESEARCH

Motivation: Emotions run high on the topic of animal testing in research, with strong positions being taken by both sides. Medical researchers claim that breakthroughs in essential areas of medicine, such as cancer treatment and organ transplants, will require the continued use of animal testing procedures. They point to past successes in which animal testing was involved in the development of the polio vaccine and of insulin for the treatment of diabetes, and ask, "Who would be willing to trust a new drug, a new medical treatment, or a new surgical procedure that had not been adequately tested and found to be both safe and effective?" Those who oppose animal testing procedures would like scientists to find and use alternative methods of testing. Some animal rights activist groups have demonstrated their opposition by staging protest demonstrations and, in some cases, by breaking into research laboratories, kidnapping experimental animals, and even planting bombs in the laboratories. The controversy calls for the benefits to man to be weighed against the claim of cruelty to animals, and for the possibility of developing and employing alternative methods of reliable testing to be explored.

Recommended Grade Level: Grades 6–8

Strategies Involved: Investigative approach
                          Student involvement
                          Student interaction
                          Decision making based on critical thinking

Materials Required:

- a collection of magazine and newspaper articles expressing viewpoints both for and against the use of animals in scientific research
- classroom and library reference materials
- a chalkboard

Procedure:

Involve all students in a class activity, following general guidelines:

1. Have individual students present to the class short reports based on newspaper and magazine articles that express viewpoints of individuals and/or organizations that:

   - support the use of animals in scientific research,
   - oppose the use of animals in scientific research.

2. Organize the class into two teams. Then, instruct each team to proceed as follows:

   - Individually and in small groups, use reference materials to gather evidence that can be used to support your team's position, FOR or AGAINST animal experimentation.
   - List in the appropriate column on a chart constructed on the chalkboard, statements that are useful for supporting your team's position.

3. Have students examine the chart and compare the FOR and AGAINST columns, with focus on:

   - evidence of benefits to man,
   - evidence of cruelty to animals,

- an assessment of the benefits versus claims of inhumane treatment of animals.

4. Have students discuss with other members of the class:

   - animal rights versus human health and well-being,
   - the need for alternative methods of testing products and procedures,
   - the dilemma of what to do in the interim if animal testing is discontinued before reliable alternative methods are developed,
   - guidelines that have been set by the National Institute of Health, the Food and Drug Administration, the Society for the Prevention of Cruelty to Animals, and the Consumer Product Safety Commission.

5. Have students discuss all the evidence presented and make a statement in answer to the question, "Should we use animals in scientific research?"

6-9 INVESTIGATING BIOTECHNOLOGY

Motivation: The introduction of new and improved consumer products, such as leaner, meatier beef and pork, as well as cholesterol-free eggs, is generally met with wide public acceptance—until it is learned that these innovations have been brought about by genetic engineering practices. Although the basic technique used is patterned after the natural tendency of bacteria to exchange genetic material, and hence give rise to new gene combinations in a cell that can then make exact copies of itself, genetic engineering is viewed with disfavor by many individuals and action groups who are opposed to any form of "tampering with nature." There are implications for the scientific community, government agencies, and independent research centers to make a full and impartial disclosure that includes both the benefits and possible risks that may accompany genetic engineering practices.

Recommended Grade Level: Grades 7–8

Strategies Involved: Investigative approach
Student interaction
Student involvement
Decision making based on critical thinking

Materials Required:

- magazine and newspaper articles dealing with benefits and risks associated with various aspects of biotechnology
- copies of student survey sheet listing WHAT IF? situations that require value-judgment decisions by individual students

Procedure:

Instruct students to proceed as follows:

1. Read the available articles expressing differing points of view about biotechnology. Focus on the reasons given by people who support the practice and by those who oppose it.

ARE YOU IN FAVOR OF BIOTECHNOLOGY?
SURVEY SHEET

	YES	NO	PASS
WHAT IF . . . you or a member of your family needs to have treatments with interferon to cure a serious illness. Knowing that interferon is produced by biotechnological techniques involving gene-splicing in bacteria, will you accept the treatment?			
WHAT IF . . . you are shopping for strawberries for decorating a strawberry shortcake. There are large berries (produced by biotechnological techniques) with little flavor and there are small berries that have the characteristic strawberry flavor. Will you choose the large berries?			
WHAT IF . . . you are concerned about excess fat and cholesterol in your diet. Knowing that "regular" pork is 25 percent fat, while that from biogenetically programmed animals is only 5 percent fat, will you decide to include the low-fat-content pork in your diet instead of the meat produced by traditional methods?			
WHAT IF . . . a genetics research organization in your state is planning to test a newly altered strain of bacteria to determine its ability to protect developing fruit from frost. Although no field tests have yet been made, the director of research offers to spray your cherry trees to prevent damage due to unseasonably cold temperatures expected to occur tonight. Will you accept the offer?			
WHAT IF . . . your doctor has prescribed carrots, which you do not like, for your nutritional needs. A new orange-colored cauliflower that contains carotene (normally found in carrots) has been developed. Will you agree to try the "new" cauliflower?			
WHAT IF . . . you wish to purchase red roses to give to your mother on Mother's Day. The florist tells you that, in addition to his naturally produced fragrant roses, he has roses that have been "engineered" to last longer but are lacking in fragrance. Will you purchase the long-lasting roses?			

2. Participate in an open class discussion:

   - Share with classmates information and findings obtained from reading and other outside sources, FOR and AGAINST the employment of a variety of procedures that involve biotechnology.
   - Express individual viewpoints and exchange ideas about the issue, using the terms biotechnology, bioengineering, and genetic engineering in a meaningful way.

3. Giving reasons to justify your decision and help others who are unable to make up their minds, respond, in writing, to a series of WHAT IF? situations that are concerned with specific examples of biotechnology. (See survey sheet.)

4. Contribute some additional WHAT IF? situations to which other members of the class can respond.

As a follow-up, have students:

1. Compile a record of responses made by individuals and determine if the class generally favors or rejects the practice of biotechnological procedures.

2. Use all collected information to prepare a class statement describing the ideal situation that should be the ultimate goal of bioscience, and indicate the place, if any, of biotechnology and genetic engineering in the overall plan.

6-10 IDENTIFYING CHEMICAL FOOD ADDITIVES

Motivation: Over the centuries man has associated food with his well-being. In early civilizations he learned to avoid poisonous mushrooms and spoiled food and to preserve his food from decay by sun-drying fruits and by salting and smoking meats. Later, explorers searched for routes to the Orient in their quest for spices that would prevent food spoilage, and, eventually, methods of canning, refrigeration, freezing, and irradiating foods were developed for retarding food spoilage. Although there are very few cases of food poisoning today, and foods have a longer shelf life than ever before, there is a growing concern that some food additives may be associated with possible health hazards. Concerned groups are calling for the scientific community to develop methods of food processing that would incorporate the maximum benefits to be derived with the minimum risk to people's health.

Recommended Grade Level: Grades 5–8

Strategies Involved: Student involvement
Investigative approach
Science skills development

Materials Required:

Each group of four students will need:

   - a collection of labels from assorted prepared and packaged foods
   - library reference materials
   - a notebook

Procedure:

Instruct students, working in groups of four, to proceed as follows:

1. Individually, survey five households and record the responses relating to purchases of "regular" or "all natural" foods.
2. Working with other members of your group, consolidate the information gathered by group members and analyze the data to determine the general attitude expressed—FOR or AGAINST food additives.
3. Examine a collection of labels from assorted canned and packaged foods:

   - Identify foods that are "all natural" and those that contain chemical additives.
   - Use references to learn the reasons why each of the additives is used and the possible risks, if any, that laboratory tests indicate may accompany their use.

4. Prepare a notebook chart that incorporates all pertinent information obtained by reference work and the examination of food labels.
5. Discuss with other group members:

   - which chemical additives appear to have widespread use,
   - if the additives used are primarily for cosmetic effects or for retarding the spoilage of food,
   - reasons why those foods labeled "all natural" are more costly than those that contain preservatives,
   - the benefits and the risks involved in using food products that contain chemical additives.

6. Prepare a group statement that tells how reading product labels can be helpful to shoppers.

6–11   REDUCING NOISE POLLUTION

Motivation: Noise pollution is as dangerous to human health as are air and water pollution—and its effects are more immediate. Whether in the form of unwanted sounds associated with our highly technological society or with current trends for greatly amplifying the sounds of modern music, the public is constantly being bombarded with excessive noise levels, many times exceeding the average pain threshold, which lies between 130 and 140 decibels.

There is a need for limiting the intensity of sound in the home, in the classroom, in the workplace, in public places, and in areas associated with major highways and airports. The employment of methods for absorbing sound waves has proven to be effective in controlling some of these unwanted noises. In both indoor and outdoor situations, noise pollution is everyone's concern.

Recommended Grade Level: Grades 6–8

Strategies Involved: Hands-on activity
Student involvement
Student interaction
Investigative approach

Materials Required:

- a tape player
- tapes of high noise levels recorded at a peak time in a factory, machine shop, or press room

In addition, each group of four students will need:

- two identical shoeboxes
- an X-acto knife or scissors
  NOTE: STUDENTS MUST BE SUPERVISED WHEN USING SHARP INSTRUMENTS.
- a marking pen
- a metric ruler
- a clicking device
- glue
- foam rubber, felt, terrycloth, carpeting, styrofoam, or similar material

Procedure:

Instruct students to proceed as follows:

1. Listen to a tape of high intensity noise for a period of about five or ten minutes.
2. Share with other members of the class the reactions you experienced to the noise.

Then, have students, working in groups of four, continue as follows:

1. Engage in an investigation with other members of your group:

- At each end of two identical shoeboxes, cut a circular hole that is 5 cm in diameter.
- Select a material with which to line one box. Then cut pieces of the selected material to fit the top, bottom, and all four sides of the box, cutting holes in the end pieces to match those in the ends of the box.

- Use glue sparingly to hold the pieces of lining in place.
- Put the lids on both boxes and mark the top of each lid to identify the lined and unlined box.
- Allow each group member to conduct a test:

    1) Place your ear against the opening at one end of the lined box while another student operates a clicking device being held just within the hole at the other end.
    2) Repeat the test for the unlined box.

2. Compare the results experienced by all group members and discuss:

- why the noise level was lower in the lined box,
- how the investigation relates to the noise level produced by a machine in a carpeted room with drapes, for example, as compared with the noise level produced by the same machine in an otherwise empty room,
- why libraries are usually carpeted,
- in what practical situations the information gained from this investigation could be useful.

3. On the basis of the knowledge gained from this investigation, write a group statement suggesting some outdoor materials that could be used to absorb excessive noise from heavy street traffic in a residential area and, on a broad base, showing how urban and regional planning boards could find this information useful.

6–12   SHARING RESPONSIBILITY FOR THE SAFE DISPOSAL OF WASTES

Motivation: Although dumping waste into nearby waterways is a convenient and long-established practice, it has become a serious problem in the modern world. Since the days of early villages and settlements, there has been an enormous increase in the kinds and amounts of waste produced by an increasingly large and crowded population and its ever-growing highly developed technology. Today, the natural environment, many forms of wildlife, and man himself are all harmed by consequences arising from excessive and toxic wastes, nonbiodegradable products, and some shortsighted solutions to problems dealing with waste removal. A solution to the problem calls for decision making that will be effective in resolving this conflict between man and nature.

Recommended Grade Level: Grades 4–8

Strategies Involved: Investigative approach
Student interaction
Critical thinking skills development

Materials Required:

- a bulletin board display focusing on problems caused by the dumping of waste in waterways

- a collection of pictures and articles from newspapers and magazines that report specific cases of waste being dumped in water and the resulting problems that are noted

Procedure:

Instruct students to proceed as follows:

1. Examine the pictures and accompanying headlines and/or captions mounted on the bulletin board. In particular, look for scenes of:

   - fish floating in water containing industrial waste,
   - birds with oil-coated feathers,
   - beaches littered with assorted hospital waste,
   - dead marine birds, strangled by plastic ring "choke collars" from six-packs,
   - dead sea turtles caught in plastic fishing nets,
   - human entanglements in lost fishing lines.

2. Add pictures of similar situations to the display.
3. Select a picture from the display indicating a problem for study.
4. Read available articles relating to the problem selected.
5. Prepare and present a report, based on findings about the problem, selected and studied on an individual basis.
6. Discuss with other members of the class:

   - the cause of the problem being investigated,
   - the seriousness of the problem,
   - alternatives that have been suggested for the alleviation of the problem.

7. Make a composite list of suggestions for possible ways to solve the problems that have been investigated by individuals and indicate which could be directed specifically to individuals, to industry, to communities, and to government agencies.

7

Using an
Integrated Approach
to Science

There is much truth in the oft-quoted observation that "you cannot do just one thing." In all probability there is no place that this truism—paraphrased, "students cannot learn just one thing"—is more dramatically demonstrated than in the science classroom. Activities that focus on *just* the science content or *just* the science skills and processes do not provide meaningful learning experiences for students in the middle grades. At this level, science activities involving other disciplines should be included as well.

A curriculum design for unified science tends to be student-centered. It focuses on objectives that bring forth a change in attitudes, allowing students to develop broad perspectives. Activities that encourage them to tackle a problem of the real world allow them the freedom to pursue an investigation as it spills over into areas of social science, language arts, mathematics, health education, physical education, and—in some cases—arts and crafts. With this approach, artificial boundaries between "separate" disciplines disappear, and students are provided with incentives for learning what they need to know in order to find the answer to a challenging problem or interesting question that served to initiate the investigative activity.

ACTIVITIES THAT USE AN INTEGRATED APPROACH TO LEARNING

7–1 DISPLAYING INTERESTING DESIGNS IN NATURE

Motivation: The designs of some of the most commonly known objects in nature give evidence of the order that exists in the natural world. Snowflakes and ice crystals found in frost are all six-sided figures, fern fronds and plant leaf specimens have an amazing symmetry of form and balance, which characterizes all members within a species, and a bird's quill feathers are equipped with tiny hooks by which the vane of a ruffled feather can be reassembled and held together to ensure its proper alignment for functioning in an effective manner. Lasting records of some of these interesting marvels of design can be prepared by students for display in the classroom or at home.

Recommended Grade Level: Grades 4–6

Strategies Involved: Hands-on activity
Student involvement
Interdisciplinary approach

Materials Required:

Each group of four students will need:

- a collection of leaves and/or fern fronds
- a stack of heavy books
- newspapers
- paper toweling
- pencils (one per student)
- food coloring or colored ink
- shallow bowls (one per color used)
- water
- heavy art paper
- straight pins
- toothbrushes (one per student)

Procedure:

Instruct students to proceed as follows:

1. Working in groups of four:
   - Select specimens of leaves or fern fronds that are of suitable size and shape for making art prints.
   - Place these specimens between layers of paper toweling and press them under a stack of heavy books.
   - Spread protective layers of newspaper over a large area on a table top that is to be used as a work area.
   - In a shallow bowl, dilute food coloring with water to make the degree of color intensity desired.

2. Working individually:

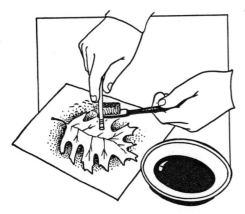

- Make a selection from the specimens that have been pressed and position them, either singly or in small group arrangements, on heavy art paper that has been placed on top of the newspaper padding. Then, pin down the tips and edges of the specimens to avoid curling and to prevent any food coloring from seeping under the specimen.
- Dip the bristles of a toothbrush in the coloring material and hold the brush over the mounted specimen. Then, rub the sides of a pencil over the wet bristles, spraying a fine mist of food coloring over all exposed sections of the paper. Control the spraying so that it is heaviest near the specimen and gradually becomes lighter as it nears the outside edges of the paper.
- After the dots of color have dried, unpin and remove the specimen.
- Examine the art paper, which shows dots of color in varying degrees of intensity everywhere except where the specimen has been pinned.

3. If time permits, repeat the procedure outlined in step 2 above to make a different print in a different color.
4. Display all completed prints on the bulletin board to be viewed for a study of interesting designs in nature.

7-2 ADAPTING A DESIGN IN NATURE TO ONE OF MAN'S INVENTIONS

Motivation: Many of man's so-called inventions are actually copies or adaptations of designs found in nature. There are, for example, similarities that can be observed in the structure and workings of a fish and a submarine, and of a dragonfly and an airplane. These links between nature and technology indicate man's ability to find inspiration for technological developments through an understanding and knowledge of the natural world.

Recommended Grade Level: Grades 5–8

Strategies Involved: Hands-on activity
Student involvement
Interdisciplinary approach
Creative expression

Materials Required:

Each group of four students will need:

- a collection of parachute-type seeds from plants, such as the milkweed and dandelion

- art paper and supplies
- at least one magnifying lens

Procedure:

Instruct students, working in groups of four, to proceed as follows:

1. With and without the aid of a magnifying lens, examine the structure of a parachute-type seed.
2. Release the seed from a height above the head and observe its descent in still, as well as in moving, air.
3. Describe the manner in which nature's parachute is carried through the air.
4. On the left side of a piece of art paper draw a diagram of a parachute-type seed, and on the right side draw a diagram showing the basic design of a parachute. Between the two diagrams, draw one or more additional diagrams showing intermediate stages by which the structure of the seed may have metamorphosed in the inventor's mind to become the basic design for a working, man-made parachute.
5. Discuss with other group members how other inventions also may have been modeled after things in nature. Consider the insect stinger and hypodermic needle, the dragonfly and airplane, and other examples.
6. Make a summary statement in which you tell how a knowledge of the natural world helps man to develop his art and his technology.

7–3 GATHERING INFORMATION ABOUT THE EARTH'S HISTORY

Motivation: Fossils reveal much information about the earth's past and about the kinds of plants and animals that were present at a particular time in the earth's history. Some unusual fossils reveal information about the structure of the woolly mammoth, the saber-toothed tiger, and a prehistoric flying reptile. More common fossils of sea animals have been found in inland streams and desert areas, revealing that these areas were once under seawater.

Dating techniques for determining the age of fossils indicate that some fossils, such as those of giant dinosaurs, were formed over one hundred million years ago. Life forms of the period so pinpointed can then be visualized in the setting of earth conditions believed to have existed at that time.

Recommended Grade Level: Grades 4–8

Strategies Involved: Student involvement
Interdisciplinary approach
Library research skills development
Creative thought synthesis

Materials Required:

- a collection of fossils or pictures of fossil formations
- classroom and library reference books
- writing materials

Procedure:

Instruct students to proceed as follows:

1. Select a fossil specimen or a picture of one that looks interesting.
2. Examine the fossil and gather as much information as possible about it:

   - location where the fossil was found,
   - identification of the imprint(s).

3. Use library reference materials to supply additional information:

   - the geologic age of the earth at the time the organism represented by the fossil is believed to have lived where the fossil was formed,
   - the earth conditions believed to have existed at the time the "fossil" organism was living an active life,
   - other forms of life believed to have existed at the same time the fossil plant or animal inhabited the earth.

4. Develop a mental picture of the earth and its inhabitants as they might have looked at the time in the earth's history indicated by the investigation.
5. Write a short story about how these inhabitants might have interacted with one another as they played out one of life's dramas, or as they experienced some event that might reasonably have occurred at that time in the earth's history.

When all student stories have been completed, assemble them in a folder, which can be kept on the classroom reading table for student recreational reading.

7–4 COMPARING LIFE SPANS OF DIFFERENT ORGANISMS

Motivation: Living a full "lifetime" does not represent the same amount of time for all organisms. Among the oldest living things on earth today are plants such as the giant sequoias and bristlecone pines that are over 1,000 years old, while some bacteria are known to live a mere 20 minutes before they split to form two new organisms. Some studies have indicated that longevity of mammals is related to their heartrates, with each species having about the same number of heartbeats during its entire life span—those having faster heartrates completing their lifetime total in a shorter time than those whose rates are slower. Longevity records for many animals can be used to compare their typical life spans with that of man.

Recommended Grade Level: Grades 6–8

Strategies Involved: Student involvement
                          Science skills development
                          Interdisciplinary approach

Materials Required:

Each student engaging in the activity will need a copy of the LIFE SPAN OF MAN AND ANIMALS WORKSHEET.

Procedure:

Instruct students to proceed as follows:

1. Examine the worksheet listing the typical life spans for several animals and note the wide range represented by various animals.
2. Compare the typical life span of the Indian elephant and man, noting that the elephant completes its life span in one-half the time that is typical for man.
3. Complete the worksheet, calculating the fractional part of man's life span that each animal needs to complete its own life span.

   HELPFUL HINT: Be sure to reduce all fractions to lowest terms.

4. On the basis of your calculations, determine what is meant by the statement, "The age of a 10-year-old dog is equivalent to that of a 60-year-old man."
5. Consider organisms that are known to live longer than man and express man's typical life span as a fractional part of:

   • the life of a tortoise whose age has been determined to be 144 years,
   • the life of a redwood tree which, when cut down, showed 3,600 annual rings in its cross-sectional view.

LIFE SPAN OF MAN AND ANIMALS WORKSHEET

ANIMAL		MAN	COMPARISON
Name	typical life span in years	typical life span in years	fractional part of man's life span required by animal to complete its entire life span
Indian elephant	36	72	$\frac{36}{72} = \frac{1}{2}$
Dog	12	72	$\frac{12}{72} =$
Horse	24		
Squirrel	9		
Rabbit	8		
Chimpanzee	18		
Guinea pig	6		
Guppy	2		

7–5 INVESTIGATING THE EFFECT OF TEMPERATURE ON THE SHELF LIFE OF DAIRY PRODUCTS

Motivation: The "sell by" date stamped on containers of milk and other dairy products indicates the last date that an item should be sold, assuming it has been stored and handled properly. Generally, the consumer is assured that the item is safe for consumption for a period of about one week after this date. Storage of dairy products at proper temperature levels is, of course, necessary both before and after their purchase. Refrigeration of milk not only enhances its taste and flavor—it also prevents the rapid growth of bacteria that causes milk to sour. An activity that focuses on the relationship between temperature and the average shelf life of milk enables students to develop an awareness of "sell by" dates and of the need to refrigerate certain foods promptly after their purchase.

Recommended Grade Level: Grades 6–8

Strategies Involved: Student involvement
Interdisciplinary approach
Science skills development

Materials Required:

Each student engaging in the activity will need:

- a "sell by" date imprint from a recently purchased container of milk, brought in from home or from the school cafeteria
- a copy of the chart RELATIONSHIP BETWEEN TEMPERATURE AND THE AVERAGE SHELF LIFE OF MILK BEYOND ITS "SELL BY" DATE
- a ruler
- graph paper

Procedure:

Instruct students to proceed as follows:

1. Examine the chart showing the relationship between temperature and the average shelf life of milk beyond its "sell by" date.
2. Prepare a graph that will provide for temperature readings, in degrees Fahrenheit, along the X-axis and average shelf life, in number of days, along the Y-axis.
3. Using information from the chart, plot values on the graph and connect these points to make a line graph.

RELATIONSHIP BETWEEN TEMPERATURE AND AVERAGE SHELF LIFE OF MILK

Temperature	Average Shelf Life	
40° F.	10	days
45° F.	5	days
50° F.	2½	days
55° F.	1	day
60° F.	less than 1	day

4. Analyze the graph and determine:

- why milk left out on a table or counter becomes sour very quickly,
- the average number of days beyond its "sell by" date that milk may be expected to be safe to drink, if refrigerated at 42° F.,
- how the graph is useful for determining the relationship between the temperature at which milk is stored and its expected shelf life beyond its "sell by" date.

5. Check the "sell by" date on your milk carton and, using the general consumer guideline that milk refrigerated at 40° F. will remain good for use up to ten days beyond its "sell by" date, determine the date beyond which milk in your carton would be considered unsafe to drink, even though kept in a refrigerator at a temperature of 40° F.

6. Write a short paragraph about the importance of refrigeration to the freshness of milk and other dairy products and explain why "sell by" rather than "use by" dates are frequently stamped on packages containing these products.

7-6 BECOMING AWARE OF THE MECHANICAL ADVANTAGE OFFERED BY THE USE OF A LEVER

Motivation: Man's progress can, in part, be attributed to his ingenuity in devising and using machines to make his work easier. In addition to the highly sophisticated machinery that has been developed for use in transportation, in industry, and in modern appliances for the home, some very simple machines—such as levers—are also used in the performance of some routine tasks in our daily lives. For example, we may make the work associated with some of our activities easier by using an appropriate "tool" in the form of a wheelbarrow, a crowbar, a nutcracker, a pair of tongs, or a claw hammer. Through the use of such tools we gain a certain mechanical advantage, which makes the performance of the task easier. When Archimedes recognized this advantage, he made his now-famous statement: "Give me a lever that is long enough and a fulcrum on which to rest it and I will move the earth."

Recommended Grade Level: Grades 7–8

Strategies Involved: Hands-on activity
Student involvement
Interdisciplinary approach
Science skills development

Materials Required:

Each group of four students will need:

- a meter stick
- a wedge-shaped block to serve as a fulcrum
- a spring scale calibrated in grams
- a 100-gram weight

- string
- a copy of the INVESTIGATING LEVERS WORKSHEET

Procedure:

Instruct students, working in groups of four, to proceed as follows:

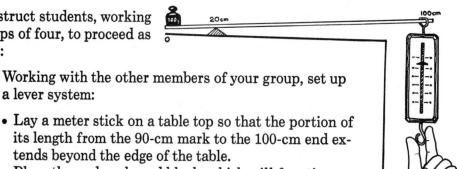

1. Working with the other members of your group, set up a lever system:

   - Lay a meter stick on a table top so that the portion of its length from the 90-cm mark to the 100-cm end extends beyond the edge of the table.
   - Place the wedge-shaped block, which will function as a fulcrum, under the 20-cm mark on the meter stick.
   - Fasten the 100-gram weight to the 0-cm end of the meter stick.
   - Attach a spring scale to the 100-cm end of the meter stick and pull down on the hook of the spring scale until the meter stick is raised at the weighted end and becomes parallel to the table top. Check the spring scale to find out how much effort was needed to lift the weight.

2. Record all information—amount of weight to be lifted, effort required to lift the weight, length of the effort arm of the lever system, length of the resistance arm of the lever system—in appropriate columns on the "levers" worksheet.

3. Calculate the mechanical advantage of the lever system, using the formula

$$\text{M.A.} = \frac{\text{length of the effort arm}}{\text{length of the resistance arm}}$$

   and record this value in the space provided on the worksheet.

4. Move the fulcrum to other locations—10-cm, 25-cm, 30-cm, and 40-cm marks on the meter stick—and again take readings, make measurements, and calculate mechanical advantage values for recording appropriately on the worksheet.

5. Analyze the data on the worksheet and, for each lever system in use:

   - determine the relationship between the relative length of the effort arm and the amount of effort required when using a lever system to lift an object,
   - note the relationship between the relative length of the effort and resistance arms and the mechanical advantage of the lever,
   - compare the mechanical advantage with the ratio

$$\frac{\text{amount of weight to be lifted}}{\text{amount of effort required to lift the weight}}$$

Name ———————————————— Date ————————————

INVESTIGATING LEVERS WORKSHEET

$$\frac{\text{Length of Effort Arm}}{\text{Length of Resistance Arm}} = \text{Mechanical Advantage}$$

Amount of weight to be lifted	Amount of effort required to lift the weight	Mechanical Advantage Calculation		
		Length of the effort arm	*Length of the resistance arm*	*Mechanical advantage*
100 g		80 cm	20 cm	$\frac{80}{20} = 4$

6. Using the formula for calculating the mechanical advantage of a lever, solve the following:

- A crowbar that is 140 cm long is used to pry up some boards. If the fulcrum is placed 20 cm from the boards, what is the mechanical advantage of the crowbar?
- A seesaw that is 10 m long is placed on a fulcrum that is 2 m from the resistance. Find the mechanical advantage of the seesaw.

HELPFUL HINT: Use the worksheet pattern to record the data and make calculations.

7–7   COMPARING "STUDENTPOWER" AND HORSEPOWER

Motivation: Automobile engines, as well as motors used in airplanes, motorboats, air conditioners, and other commercial and industrial appliances, are rated on the basis of their performance capabilities. These are expressed in terms of *horsepower,* a unit originally based on findings that a horse can do 550 foot-pounds of work in one second, but currently applied to the amount of work that can be done by a working system in a specified period of time.

An activity that investigates student performance of a task allows students to compare "studentpower" with "horsepower" and provides both purpose and motivation for developing a high degree of proficiency in their basic math skills.

Recommended Grade Level: Grades 6–8

Strategies Involved: Hands-on activity
Student involvement
Student interaction
Interdisciplinary approach
Science skills development

Materials Required:

Each student/partner combination will need:

- a tape measure or ruler
- a stopwatch
- two copies of the student worksheet (one per student)

In addition they must have access to

- a bathroom scale,
- a flight of stairs, and
- the master record chart on the chalkboard.

Procedure:

Instruct students, working with partners, to proceed as follows:

1. Working with your partner, measure the vertical height, in feet, of a flight of stairs to be used in the activity and record the information on the master chart on the chalkboard.

> HELPFUL HINT: Measure the height of one step and multiply by the number of steps.

2. Determine your weight, in pounds, and record this information on the master chart.

3. Run swiftly up the stairs while your partner uses a stopwatch to time your performance, in seconds.

4. Use the information collected to calculate how much horsepower you developed to accomplish the work involved in "lifting" your body from floor level to the top of the stairs in the amount of time recorded. Use the formula that is based on James Watt's findings that "a good horse could lift 550 pounds of coal a distance of 1 foot in 1 second":

$$\frac{(\text{weight in pounds}) \times (\text{height in feet})}{550 \times (\text{time in seconds})} = \text{Horsepower}$$

5. Enter all data in the appropriate columns of the chart on the chalkboard.

6. When you are sure that all students in the class have recorded their individual data for the activity, transfer all information from the master chart on the chalkboard to your worksheet and analyze the data:

   - What information was needed to calculate the horsepower generated?
   - Does the record indicate that most students develop more or less than 1 horsepower?

7. Discuss with your partner the horsepower developed by various members of the class:

   - Are the students who developed the greatest amount of horsepower also members of a sports team?
   - How would the amount of horsepower generated be affected by the requirement that each student carry a 20-pound box of books while running up the stairs?
   - Could the horsepower generated by each student be maintained at the same rate if the stairway were extended to the top of a 25-floor building?

8. Make a list of devices whose power is calibrated in horsepower. Then, indicate the advantages of using these motorized appliances instead of human power or the power of a horse.

Name _____ Date _____

POWER DEVELOPED BY STUDENTS
WORKSHEET

$$\text{Horsepower} = \frac{(\text{weight in pounds}) \times (\text{height of stairs in feet})}{550 \times (\text{time in seconds})}$$

Name of student	Weight	Height of stairway	Time required	Horsepower developed

7–8   CALCULATING RELATIVE HUMIDITY

Motivation: We generally explain our discomfort on a hot, sticky day as being due, not to the heat, but to the high humidity. On some hot days the heat is more bearable because of a lower humidity level. From these experiences we know that the humidity—the amount of water vapor in the air—can vary, and that the water vapor in the air is not always near its saturation point. Ratios between the amount of water vapor in the air, compared to the amount that can be held at its saturation point at the same temperature, can be calculated, with the resulting percentage referred to as "relative humidity."

Recommended Grade Level: Grades 6–8

Strategies Involved: Hands-on activity
Student involvement
Science skills development
Integrated learning approach

Materials Required:

Each group of four students will need:

- four small, transparent plastic boxes*, 1 by 1 by ¾ in. (one per student)
- 100 small lead shot pellets (25 per student)
- four student worksheets (one per student)

Procedure:

Instruct students, working in groups of four, to proceed as follows:

1. Set up a small transparent plastic box to represent a classroom containing air at a temperature that holds 25 parts of water vapor when saturated. Place 25 pellets of lead shot, each representing one part of water vapor, in the box.
2. Distribute the pellets as evenly as possible inside the box and calculate the relative humidity, using the formula:

$$\text{Relative Humidity} = \frac{\text{parts of water vapor present}}{\text{parts of water vapor present at the point of saturation}}$$

   Then record the data pertaining to this situation on the worksheet.
3. Remove five pellets from the box and calculate the relative humidity for this situation, recording data, as before.
4. Continue removing pellets from the box, as specified on the worksheet, and record appropriate data for each condition of relative humidity.
5. Analyze the data collected and complete the worksheet.
6. Discuss with other members of your group:

   - what happens to the relative humidity as the number of units of water vapor is reduced,

*Available from Carolina Biological Supply Co., Burlington, N.C. 27215.

RELATIVE HUMIDITY OF AIR AT A GIVEN TEMPERATURE
WORKSHEET

$$\text{Relative Humidity} = \frac{\text{parts of water vapor present}}{\text{parts of water vapor present at the point of saturation}}$$

	parts of water vapor present	maximum number of parts of water vapor that can be held at given temperature	Relative Humidity Calculation
	25	25	$\frac{25}{25} = 100\%$
	20	25	
	15		

- what personal experiences you can recall that suggest a relationship between temperature and relative humidity,
- what you would expect to happen to the relative humidity in your "room" model containing 15 parts of water vapor if the temperature were raised, if lowered,
- under what conditions you would recommend the use of a room humidifier, of a dehumidifier,
- how the humidity level at different temperature levels contributes to a feeling of physical comfort or discomfort.

7–9 DESCRIBING DETAILS OF AN OBSERVATION TO OTHERS

Motivation: Action-based science studies of all kinds heighten the joy students derive from watching something in motion, while at the same time, they provide an incentive for the students to become acute observers. The curiosity and sense of wonder that students exhibit whenever presented with a device that moves continuously can be engaged and focused on the workings of a simple object, such as a drinking bird model. After viewing this intriguing novelty for only a short time, students can communicate their observations, both orally and in writing. Describing accurately what they observe helps them develop an understanding of a science topic, which they can then share with others.

Recommended Grade Level: Grades 7–8

Strategies Involved: Hands-on activity
Student involvement
Student interaction
Science skills development
Interdisciplinary approach
Reinforcement of learning

Materials Required:

Each group of four students will need:

- a drinking bird model*
- a glass of water
- classroom and library reference materials

Procedure:

Instruct students, working in groups of four, to proceed as follows:

1. Place a glass that is almost full of water on a table top where it can be viewed clearly by all group members.
2. Carefully place the drinking bird model next to the glass.
3. Gently submerge the head of the bird in the water.

*May be ordered from FREY Scientific Company, Mansfield, Ohio 44905.

4. Observe what happens, noting particularly any changes in the level of the colored liquid in the tube that serves as the bird's body. Repeat the process one or more times.

5. Write a detailed description of the bird's activities.

6. Read the description written by another group member while he/she reads yours in order to discover points of agreement and disagreement about what was observed. Then, set the model in motion one more time to check the accuracy and completeness of the descriptions written.

7. Discuss with other members of the group:

- how the drinking bird model works, recalling discoveries made in Activity 2–9 and using reference materials for help in preparing an explanation that can be verified,
- how an accurate and complete description of observations made contributes to an understanding of how the model works.

7–10 INVESTIGATING HOW SMOKING CAN BE HAZARDOUS TO THE HEALTH OF NONSMOKERS

Motivation: It has been known for some time that smoking is a serious health hazard. The World Health Organization reports that each year tobacco-related diseases, such as cancer, emphysema, and heart disease, take a heavy toll of 2.5 million lives. Although the relationship between these diseases and smokers has been well-established, recent findings indicate that nonsmokers also are victims of health problems caused by smoking. By breathing air polluted by tobacco smoke in the home or in public places that allow smoking, these "involuntary smokers" are exposed to many of the same health hazards as are active smokers.

Recommended Grade Level: Grades 4–6

Strategies Involved: Hands-on activity
Student involvement
Student interaction
Science skills development
Integrated learning approach

Materials Required:

Each group of four students will need:

- an unsmoked cigarette
- an ashtray or metal jar top to serve as an ashtray

- two tripods
- two glass plates
- matches

Procedure:

Instruct students, working in groups
of four, to proceed as follows:

1. On widely separated sections of a
   work table in a well-ventilated
   room, set up two tripod supports,
   with a glass plate resting on the
   ring top of each. Below one tripod,
   place an ashtray with an unlit
   cigarette resting on it.
2. Designate one student to strike a match and light the cigarette, as it remains in its position on the ashtray, so that the smoke rises to the glass plate
   above. Shield the side, or make other necessary adjustments to ensure that
   the smoke travels upward.
3. Allow the entire cigarette to be "smoked" before disturbing the setup and
   discarding the ash collected in the ashtray.
4. Remove the glass plates from both tripods and compare their lower surfaces, noting the accumulated residues from the cigarette smoke.
5. Discuss the investigation with other group members:

   - Give reasons for the inclusion of the second tripod/glass plate in this
     experimental design.
   - Identify the dangers to nonsmokers who breathe the air in an environment
     polluted by smokers.
   - Check out some arguments that have been offered FOR and AGAINST
     the restriction of smoking on trains, buses, and airplanes and in homes,
     restaurants, and many public places.
   - Suggest some ways to resolve the conflict that is associated with the issue
     of the rights of smokers versus the rights of nonsmokers.

7–11 DISCOVERING THE IMPORTANCE OF WATER TO LIFE

Motivation: All living things need water. There are some forms that cannot survive outside a watery environment, and none can maintain its structure or engage
in normal life processes without supplies of safe, usable water in sufficient
amounts. Of course, not all living material contains the same amount of water.
While the average percentage of water in living material is approximately 80 percent, some organisms, such as jellyfish, have percentages that are considerably
higher. Any reduction in the percent of water content in the structure of organisms
can be observed in the form of weight loss, due to dehydration. Some plants, when
deprived of water, begin to wilt, then droop; in many cases, death follows in a

matter of days. The percent of water in a variety of plant tissues can be calculated, based on data collected in a student activity.

Recommended Grade Level: Grades 6–8

Strategies Involved: Hands-on activity
                         Student involvement
                         Science skills development
                         Integrated learning approach

Materials Required:

Each group of four students will need:

- four different specimens of plant origin (tomato, cucumber, apple, carrot, celery, or other)
- a weighing balance
- a shallow tray
- paper toweling
- graph paper
- marking pens in four colors
- a ruler
- a student-prepared chart for recording data collected
- classroom and library reference materials

Procedure:

Instruct students, working in groups of four, to proceed as follows:

1. Conduct an investigation:

- Line a shallow tray with several thicknesses of paper toweling.
- Select four different plant materials that will fit in well-separated positions on the tray.
- Prepare a chart that provides for making records of daily weights taken over a period of two weeks for each material.
- Weigh each specimen, individually, and record its weight on the prepared chart.
- Place the specimens in well-separated positions on the toweling in the tray. Then, place the tray in a protected area of the classroom.
- Each school day, weigh each specimen and record its weight appropriately on the chart. Replace the toweling as needed.

2. At the end of a two-week period, or when the materials are dry, use the collected data to prepare a graph:

- Plan a graph scale that will accommodate all weights represented and the total number of days involved in the investigation.
- Using a different color to represent each specimen, plot the daily weights and connect the points with a line to indicate the individual weight values over the time period of the investigation.

3. Compare the four line graphs drawn on the same axes and discuss with other members of the group:

- basic similarities in the general direction of all line graphs involved,
- time intervals that show rapid changes in weight,
- time intervals that show gradual changes in weight,
- how the overall weight change can be explained.

4. Use the formula

$$\frac{\text{weight loss}}{\text{original weight of specimen}} = \text{original percent of water weight}$$

to calculate the percent of water normally found in the tissues of each specimen, and write these values on the corresponding line graphs.

5. Use references to gain information about the water content of a variety of animals, such as the earthworm, jellyfish, and camel, and of the human body.

6. Using information researched and the actual weight of individual group members, calculate the water weight in each person's body.

7. Review the activity and write a short summary telling how information gained through its performance can be useful in explaining:

- a sick person's weight loss due to dehydration,
- the withered appearance of a potted plant in the classroom after a long school vacation period,
- why it is believed that life, as we know it, does not exist on the moon.

7–12 SEEKING SATISFACTORY SOLUTIONS TO PROBLEMS ASSOCIATED WITH WASTE DISPOSAL

Motivation: Throughout the early periods of human history, people were not much concerned about the pollution that they caused, for they did not feel threatened by the harmful consequences of their actions. Today, however, pollution is recognized as a major problem which is being compounded by the rapid growth in the human population. In addition, the technological advances which have been made have resulted in an alarming increase in the production of hazardous and nonbiodegradable wastes. New sites for landfills in areas previously spared and sites for incinerators designed to destroy certain chemicals and toxic materials are being sought, with little success or public acceptance. Because of the widespread nature of the NIMBY (NOT IN MY BACKYARD) syndrome, there is a need to consider all factors involved before selecting waste disposal sites, and there is also a need to explore alternative methods for dealing with the problem of waste.

Recommended Grade Level: Grades 7–8

Strategies Involved: Student involvement
Student interaction
Science skills development
Interdisciplinary approach

Materials Required:

Each group of four students will need:

- a card stating a waste-related situation or problem of concern to society
- selected reference materials, including current magazine articles, brochures, newspaper clippings, and media documentaries

Procedure:

Instruct students, working in groups of four, to proceed as follows:

1. Identify the situation stated on your group's card as a serious problem of waste disposal.
2. Use available reference materials to gather information concerning possible solutions to the problem. Consider all possibilities—dumping at sea; incineration; burial in ponds, pits, landfills, and lagoons in sparsely populated areas; burial in industrial areas already heavily polluted; exporting by barge and/ 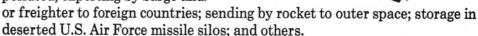 or freighter to foreign countries; sending by rocket to outer space; storage in deserted U.S. Air Force missile silos; and others.
3. Share all information gathered with other members of the group and discuss the collected findings, noting benefits and hazards, moral and ethical considerations, and concern for the environment and people's wishes involved in each, as well as the recommendations of some agencies, such as the National Institute of Environmental Health Science, the Occupational Safety and Health Administration, and the Consumer Product Safety Commission.
4. Consider a plan of action that offers the maximum benefit with the minimum risk. Then, prepare a group position paper for one member of the group to present to the entire class.

As a follow-up, have each group representative present his/her group's position paper to the class. Then, on the basis of all reports presented, have the entire class discuss the relative merits of the proposed solutions, with a special focus on the social aspects of each problem and its solution—including responsibilities for individuals, the scientific community, special interest groups and organizations, and established government agencies.

SUGGESTED WASTE-RELATED PROBLEM SITUATION CARDS
FOR GROUP CONSIDERATION

A factory uses large amounts of formaldehyde in processing its products. How should this chemical waste be handled?	A textile mill located along a waterway treats some fabrics with mercury. How should the mill dispose of the poisonous mercury after it has been used in the treatment?
A hospital accumulates waste from medical and surgical treatments given to patients. What is the best method for handling this waste?	Radon-contaminated soil has been collected from a large residential area located in the eastern part of the United States. What is the best way to dispose of this contaminated soil?
A nuclear power plant uses large drums for the temporary storage of its nuclear wastes. What is the safest method for long-range storage of these drums?	A large city collects large amounts of assorted solid waste each day. What method would be best for disposing of this waste?

8

Personal
Aspects of Science

At all levels, students show greater enthusiasm for science when they can relate to it in a personal way. Investigations that focus attention on the students themselves help to generate this enthusiasm by making their involvement highly personalized, while at the same time promoting the development of the scientific method of inquiry.

Personal concerns with grooming, health, physical fitness, and longevity are, of course, reflected in the growing popularity of health spas, home video exercise programs, and diet regimens that are subscribed to by an increasing number of adult Americans. Instead of producing effective role models for young people, however, this trend has acted to widen the generation gap—as we see growing numbers of young people becoming couch potatoes, spending more time watching television, eating junk food, and becoming overweight.

Engaging in activities that feature students as the "stars" involved in friendly competitions, comparisons, and mild tests of physical strength and endurance, at their own level, will provide an incentive for them to learn more about the specific ways human body structures are uniquely suited to perform their functions. Hopefully, this involvement will also help to draw young people into the mainstream of a health-and-fitness oriented society, resulting in an increase in the percentage of U.S. children who can pass a basic physical fitness test.

ACTIVITIES THAT FOCUS ON BODY STRUCTURES

8–1 MAKING IMPRINTS OF A NATURAL STRUCTURE DESIGNED TO SUPPORT GREAT WEIGHT

Motivation: When standing in an upright position, humans must support their entire body weight on their feet. This is made possible by the structure of the foot, which includes five main bones called metatarsals. Located between the ankle and toe bones, the metatarsals distribute the body weight most effectively when they form a natural arch that has great strength and facilitates a graceful stride. Any lowering of this arch that causes the entire sole of the foot to rest on the floor results in a condition that is commonly referred to as "flat-footedness."

Recommended Grade Level: Grades 4–8

Strategies Involved: Hands-on activity
Personal involvement
Student interaction
Investigative/discovery approach

Materials Required:

- sheets of brown wrapping paper
- colored marking pencils
- water
- classroom and library reference materials

Procedure:

Instruct students to proceed as follows:

1. Working at home:

    - Place brown wrapping paper on the bathroom floor.
    - After a bath or shower, shake any excess water from the feet and step on the paper.
    - Step off the paper and, using a colored marking pencil, outline the spots and areas on the paper that have become wet due to footprints.
    - Allow the paper to dry.

2. Bring the footprints to school on the specified date.

Instruct students, working in groups of four, to continue with the activity:

1. Use the footprints prepared at home for an analysis and comparative study.

- Identify and label the outlined areas representing individual toes, balls of feet, and heels.
- Locate and identify the region of the arch, noting examples of "high" arches among the footprints prepared by group members.

2. Look up information on the use of the arch in architecture in various construction situations and discuss with other group members:

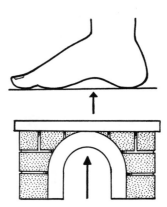

- how the arch serves to support great weights,
- how the arch of the foot may have served as a model for man's early design of arches used in doorways, aqueducts, and bridges.

3. Write a short group report that focuses on the effectiveness of the arch as a supporting structure and illustrates that "nature invented it first."

8-2 EXPERIENCING A LOSS OF BODY BALANCE

Motivation: The point at which an object will balance is called its center of gravity. It can be located for a ruler being balanced lengthwise on a narrow ledge, for a teeter-totter on a playground, or for a human body. The center of gravity for a human standing in an upright position, for example, is located in the back of the pelvis, in line with the vertebral column and the feet. However, the performance of various life activities may cause shifts in the distribution of body parts, taking them too far from the center of gravity and necessitating constant adjustments to withstand the pull of gravity. An inability to make these necessary adjustments will cause the body to suffer a loss of balance and, in some cases, to topple over.

Recommended Grade Level: Grades 7–8

Strategies Involved: Hands-on activity
                          Personal involvement
                          Student interaction
                          Inquiry/discovery approach

Materials Required:

Each student/partner combination will need:

- a ruler
- access to an uncluttered wall area

Procedure:

Instruct students, working with partners, to proceed as follows:

1. Work with your partner to perform the following:

   - Place a ruler on the floor in front of your partner and ask him/her to pick it up, without bending the knees or moving the feet.

      NOTE: More than one attempt may be necessary before he/she performs the task successfully.

      - Next, ask your partner to stand against a wall, with the back of his/her heels touching against the wall surface.
      - Place the ruler on the floor in front of his/her feet and, again, ask him/her to bend forward and pick up the ruler without bending the knees or moving the feet.
      - Observe the difficulty encountered.

2. Change places with your partner and repeat the procedure outlined in step 1, this time with you engaged in bending your body forward.

3. Discuss with your partner the observations made and the difficulties experienced:

      - the success and/or failure experienced in performing the task under the two different conditions,
      - changes in body alignment that accompanied a successful performance of the task,
      - inability to maintain balance when the body is restricted, by its position against the wall, from making adjustments to the center of gravity.

4. Make a list of specific adjustments that are made by individuals (such as circus performers walking a high wire) who must correct the center of gravity in order to maintain their balance.

8–3 DISCOVERING WHICH SIDE OF THE BODY IS "DOMINANT"

Motivation: A striking feature of the functioning of the human brain is the inverse relationship of the two cerebral hemispheres to the two sides of the body. Studies indicate that many activities performed by the left side of the body are controlled by the right side of the brain and that, correspondingly, activities associated with the right side of the body are controlled by the left side of the brain. In some cases, when one of the hemispheres is damaged, the other one takes on some of its work. Much of our knowledge about brain function has come from studying diseased and injured brains. The two hemispheres seem to be mirror images of each other, but it is believed that there must be chemical or structural differences between them that would account for innate right- and left-handedness. For most individuals a distinct dominance of one side of the brain can be identified.

Recommended Grade Level: Grades 4–6

Strategies Involved: Hands-on activity
                    Personal involvement
                    Student interaction
                    Science skills development

Materials Required:

Each student/partner combination will need:

- two copies of a "left-side/right-side dominance" task card/worksheet
- classroom and library reference materials

Procedure:

Instruct students, working with partners, to proceed as follows:

1. Record on your partner's worksheet the manner in which he/she makes spontaneous "left" or "right" responses to commands to perform each of the tasks listed on the worksheet.
2. Change places with your partner while he/she uses your worksheet for recording your responses to the same series of commands.
3. Analyze the data recorded on your individual worksheet and determine which side of your body was used more frequently in making the responses.
4. Compare individual results with your partner and discuss:

   - your body dominance and that of your partner, as determined by the greater number of responses made by either the left or the right side,
   - the influence of the left side of the brain in controlling activities performed by the right side of the body, and vice versa, as indicated in reference materials consulted,
   - the indicated brain dominance for each partner.

5. Relating this activity to personal experiences:

   - Suggest some advantages that accompany an ambidextrous condition in which a person can perform a task with equal ease, using either his/her left or right hand.
   - Give reasons why some stores sell left-handed tools and instruments such as left-handed scissors, in addition to the regular ones designed for right-handed people.
   - Determine if the pencil sharpener in your classroom is designed primarily for students who are left-handed or right-handed and give reasons for your answer.

LEFT SIDE/RIGHT SIDE DOMINANCE
PERSONAL ANALYSIS

Name of person being tested _____

Recorder _____

COMMAND	RESPONSE	
	Left Side	Right Side
Wave to a friend		
Pick up a pencil		
Write your name on the chalkboard		
Look through a rolled paper "telescope"		
Remove one shoe		
Point to the clock		
Place one foot forward		
Place your ear close to a faint sound		
Toss a ball		
Kick a wad of paper on the floor		
Carry a small suitcase		
Tilt your head to one side		

8–4   DETECTING A BODY WASTE PRODUCT RELEASED IN EXHALED BREATH

Motivation: Water is essential for life; without fresh supplies we can live only a few short days, after which the body begins to dehydrate. The importance of water is seen in the many ways it is used: for involvement in chemical processes occurring in the body, for transporting materials to locations in the body where needed, and for the removal of body wastes, such as urine from the kidneys, perspiration through the pores of the skin, and water vapor released with each breath of air we exhale from the lungs.

Recommended Grade Level: Grades 4–6

Strategies Involved: Hands-on activity
                        Personal involvement
                        Student interaction
                        Science skills development

Materials Required:

Each group of four students will need:

- four transparent plastic food storage bags (one per student)
- four strips of cobalt chloride paper (one per student)
- four small forceps (one per student)
- four medicine droppers (one per student)
- water

Additional cobalt chloride paper strips will be needed for the performance of preliminary and follow-up tests.

Procedure:

Instruct students, working in groups of four, to proceed as follows:

1. Working individually, within your group, perform the following:

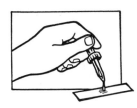

   - Examine a dry strip of cobalt chloride paper and note its color.
   - Place a drop of water on one end of the paper strip and note the color change that occurs.
   - Using forceps to handle the cobalt chloride paper, transfer one dry strip to a transparent plastic food storage bag that contains no moisture.
   - Gather the open edges of the plastic bag and hold the small opening close to your mouth while you exhale your breath into the bag.

- Quickly close the bag completely and examine the color of the enclosed strip of cobalt chloride paper.

2. Compare your results with those of other members of your group and discuss:

   - the cause of the color change in the cobalt chloride paper,
   - the source of the moisture that causes the paper strip in the dry plastic bag to change color,
   - the expected result if a cobalt chloride paper strip were to be handled by sweaty hands,
   - the source of the moisture associated with sweaty hands,
   - the reason for handling cobalt chloride paper strips with forceps.

3. On the basis of the investigation, list two ways that waste water is released from the body.

8–5 MEASURING THE NEAR POINT OF VISION

Motivation: The normal human eye has the remarkable ability to receive impressions of both faraway objects and objects close at hand. The accommodation is made spontaneously and automatically by changes in the shape of the lens of the eye as the eye shifts focus to bring the image of an object, from whatever distance, into focus on the retina.

A person with normal vision can easily observe objects at great distances, such as airplanes high above the earth's surface, a lighthouse beacon on a far-off island, or a spire atop a towering skyscraper. On the other hand, it may take much effort to "accommodate" objects at close range, since the lens must provide much greater bending power for light rays from nearby objects to come together. The closest point at which an eye can focus—its *near point*—varies with individual eyes and is often associated with age and with certain eye defects for which some "correction" may be necessary.

Recommended Grade Level: Grades 6–8

Strategies Involved: Hands-on activity
                             Personal involvement
                             Science skills development
                             Inquiry/discovery approach

Materials Required:

Each student/partner combination will need:

- an eraser-topped pencil
- a straight pin with a colored head
- a metric measuring tape
- a table of normal near point of vision values for several age groups

Procedure:

Instruct students, working with partners, to proceed as follows:

1. Mount a straight pin on an eraser-topped pencil and hold the pencil in an upright position at arm's length in front of you, while you focus on the head of the pin.
2. Gradually, move the pencil closer to your eyes, while continuing to maintain your focus on the head of the pin.
3. Continue to bring the pencil still closer to your eyes until the head of the pin can no longer be seen in sharp focus.
4. Hold the pencil at this distance—the *near point*—while your partner measures the distance in centimeters between your eyes and the pin.
5. Record this distance, and change places with your partner so that you can determine his/her near point of vision.
6. Compare the near point values determined in your investigation with those listed on the chart as normal near point of vision values for students of your age.

NORMAL VALUES FOR THE NEAR POINT OF VISION AT VARIOUS AGES										
Age in years	10	12	14	16	18	20	30	40	50	60
Near point in cm	7.0	7.3	7.6	7.9	8.2	8.5	11.2	17	50	80

7. Interpret your findings in terms of the ability of your eye lenses to change their shape and bend, as needed, to maintain the focus of an object as it approaches the eye.
8. Consult the chart and determine what happens to the ability of eye lenses to make these accommodations as a person grows older.
9. Compare the eye lens with the magnifying lens used in Activity 4–5, where a clear focus was produced only at a given, unchanging distance.

As a follow-up, have all students participate in a class discussion that focuses on how the near point of vision relates to many real life situations:

- evidence of an older person's inability to see clearly objects that are close to the eyes,
- an older person's need for eyeglasses with bifocal corrective lenses,
- observations of the eye distance people tend to maintain when reading a book, threading a needle, or when conversing with a friend,

- the tendency of some people to squint or to suffer from headaches when they engage in tedious close work such as repairing watches, working with very small beads, or making miniature models.

8-6   INVESTIGATING THE LOCALIZATION OF SOUND

Motivation: Most of us, at one time or another, have probably fumbled in a darkened room while trying to respond to the insistent ringing of a telephone or to silence a signaling alarm clock. On such occasions our ears assist us in locating the sound-emitting objects, even in the dark. Sound can be useful in other ways as well. The wail of a siren, the sharp blast of a train whistle, and the honking of an automobile horn are all used to warn us of unseen dangers. When judging the direction from which sounds emanate, binaural perception is involved. Unless the source of the sound is directly in front of us, the sound vibrations reach one ear before they reach the other so that certain accommodations must be made. The judgment of direction is less accurate for sounds above or behind us than for those directly in front of us or on one side.

Recommended Grade Level: Grades 6–8

Strategies Involved: Hands-on activity
Personal involvement
Science skills development
Inquiry/discovery approach

Materials Required:

Each student/partner combination will need:

- a watch that ticks or a clicking device
- a blindfold
- cotton earplugs
- a student-prepared chart for recording data collected

Procedure:

Instruct students, working with partners, to proceed as follows:

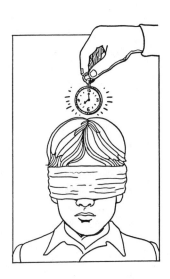

1. In a quiet room, blindfold your partner and have him/her place a cotton plug in one ear.
2. Hold a ticking watch or mechanical clicking device above your partner's head and ask him/her to report what he/she hears and the location of the source of the sound.
3. Record in chart form the actual location of the sound-producing device and the response made by your partner.
4. Then, move the watch to other locations—in front of his/her head and in back, to his/her left

side and right side, and at various angles—each time testing his/her ability to locate the source of the sound and recording on the chart both the actual location and the response made.

5. In a similar manner, but presented in a different sequence, use the same positions of the watch to test your partner's other ear.

6. Finally, perform the test with both of your partner's ears unplugged.

7. Change positions with your partner, this time with you as the experimental subject and with your partner conducting the tests and recording your responses on a chart that is similar to the one prepared for recording your partner's performance.

8. Compare the results of the tests performed and discuss some advantages of having two ears in certain situations, such as:

   • when sounds alone must be relied on to locate a ringing telephone,
   • when you are trying to locate a coin that was dropped on the pavement or an uncarpeted floor.

9. As a follow-up, plan an activity in which you and your partner can conduct an investigation to determine if it is true that "two *eyes* are better than one."

ACTIVITIES THAT FOCUS ON PHYSICAL FITNESS AND WELL-BEING

8-7 ANALYZING MUSCLE FATIGUE

Motivation: Muscle cells normally rely on aerobic respiration for their energy needs. However, even when deprived of their customary supply of oxygen, they are capable of functioning for a short time. During a period of intense or prolonged physical activity, when oxygen is being used at a faster rate than it can be supplied by the lungs, heart, and blood vessels, the muscle cells continue to function by using energy obtained via an auxiliary, anaerobic form of respiration. In the process, there is a buildup of lactic acid, causing a condition of muscle fatigue, which gradually reduces the ability of the muscle cells to continue to function. A period of rest is then needed, during which time the oxygen debt must be paid back in order to restore the muscle cells to their normal condition.

Recommended Grade Level: Grades 7–8

Strategies Involved: Hands-on activity
Personal involvement
Inquiry/discovery approach
Science skills development

Materials Required:

Each student engaging in the activity will need:

• a spring-type clothespin
• a stopwatch or watch with a sweep second hand
• a notebook for recording data collected

Procedure:

Instruct students to proceed as follows:

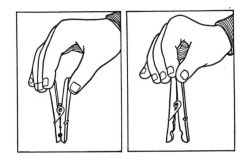

1. Pick up the clothespin and hold the ends between the thumb and index finger of your right hand.
2. Open the clothespin by squeezing the ends together; then release the pressure, allowing the clothespin to close.
3. Now, with the aid of a watch with a sweep second hand, count the number of times you can open and close the clothespin in a period of 1 minute.

> NOTE: Be sure to open the clothespin completely each time you squeeze the ends together and allow the pin to close completely between the squeeze operations.

4. Repeat the 1-minute operation, each time recording the number of times the clothespin was successfully opened and closed. Then, after ten trials with the right hand, perform the same procedure with the thumb and index finger of the left hand.
5. Examine the records for the 1-minute tests for each hand and account for any differences noted between the number of successful openings in the last 1-minute test as compared with the first test for each hand.
6. Compare the results for each hand and determine if the differences seem to correlate in any way with your "handedness" when you engage in activities such as writing, throwing a ball, or hammering a nail.
7. After analyzing and interpreting the data, repeat the procedure outlined in steps 1–3 to determine the number of times you can open and close the clothespin after a brief rest period.
8. Plan and conduct further tests to determine how much time is required for the hand and finger muscles to recover completely from the fatigue experienced after a 10-minute exercise.
9. Design an activity that investigates the fatigue of muscles in the arms, legs, back, and/or shoulders and relates the performance of individuals on the test to activities in which they normally engage.

8–8 MEASURING VITAL LUNG CAPACITY

Motivation: Although a certain amount of air must remain in our lungs to prevent their collapse, we rarely make full use of the capacity of our lungs to bring fresh air supplies into our bodies with every breath we take. Individuals with good respiratory health and strong diaphragm muscles, however, come close to achieving the maximum. Every cell in their bodies is ensured a constant supply of oxygen for use in cell respiration, during which process energy is released for engaging in metabolic activities.

There are many factors that affect the lung capacity of an individual during his/her lifetime. For example, a newborn baby, whose air exchange is less than 1 liter per minute, may increase his/her lung capacity to as much as 5 liters by the senior year in high school—if he/she becomes a tall, teenaged athlete in top physical condition. A person's lung capacity can be determined with the use of a spirometer and, allowing for individual differences such as age, height, and sex, used as an indicator of his/her physical fitness.

Recommended Grade Level: Grades 5–8

Strategies Involved: Hands-on activity
Personal involvement
Student interaction
Science skills development

Materials Required:

Each group of four students will need:

- a 1-gallon cider jug
- a large dishpan
- a 30-inch length of airline tubing
- a measuring cup, calibrated in ml
- water
- a marking pen or pencil
- a student-prepared chart for recording results of tests performed by group members

Procedure:

Instruct students, working in groups of four, to proceed as follows:

1. Pour water into a dishpan to a depth of 3 inches.
2. Fill the cider jug completely with water.
3. Place the cap on the jug. Then invert the jug and stand it in an upside-down position in the pan of water.
4. Carefully remove the cap from the jug, making certain that no water escapes from the jug and that no air enters the jug.
5. With the open end of the jug still submerged, tip the jug slightly and insert one end of the airline tube through the open end and into the neck of the jug. Check again to be sure that no air enters the jug.
6. While one group member holds the jug securely to maintain its position in the pan, exhale air in one forced breath from your lungs into the tube.
7. Examine the water level in the jug and, using a glass marking pen or pencil, mark this level on the outside of the jug.

8. Empty the water out of the jug and place it in an upright position on the table top.
9. Using a measuring cup, measure the amount of water needed to fill the jug up to the level indicated by your mark.
10. Record this volume of water, in ml, on a chart that provides for all members of your group.
11. Repeat the test for all members of the group, recording the volume of water displaced by each individual.
12. Examine the recorded information and discuss with other members of your group:

   • how the volume of water being replaced in the jug relates to the vital lung capacity of individuals,
   • differences noted in the results of performance by individuals,
   • differences noted in performance of boys and girls,
   • differences noted in performance of students of different heights,
   • the advantages of a large vital lung capacity for swimmers, divers, runners, and other athletes,
   • the relationship between a person's vital lung capacity and his/her physical fitness.

8–9   DETERMINING FLEXIBILITY

Motivation: The ease with which people move is in large part due to the flexibility of their muscles. This is recognized by athletes, who usually engage in warm-up exercises before they participate in gymnastics, tennis, skating, and other sports events. The body heat generated in the preliminaries makes their bodies more limber and helps them to perform at their best. Of course, as with other aspects of physical fitness, the flexibility of individuals varies with the extent to which their muscles have been stretched through exercise and physical activity.

Recommended Grade Level: Grades 6–8

Strategies Involved: Hands-on activity
Personal involvement
Student interaction
Science skills development
Investigative/discovery approach

Materials Required:

Each group of four students will need:

   • two 15-cm measuring tapes
   • a ruler
   • masking tape
   • a clear floor space
   • a colored pen or pencil
   • a student-prepared group record sheet

Procedure:

Instruct students, working in groups of four, to proceed as follows:

1. On a clear area of the floor, fasten a 30-cm length of masking tape. Then, using a ruler and colored pen or pencil, draw a line from end to end, down the middle of the tape. The colored line is the toe-touch line.
2. At right angles to the tape strip, place two 15-cm strips of measuring tape so that they can be used to measure distances that fall short of, as well as extend beyond the toe-touch line.
3. While wearing loose-fitting clothing that will not interfere with your body's freedom of movement, engage in a few warm-up exercises, such as swinging your arms above your head and then to your sides while jumping in place.
4. Position yourself in a sitting position on the floor, with legs outstretched and heels just touching the toe-touch line, near the 0 reading of the two measuring tapes, which extend in opposite directions.
5. Designate one member of the group to check your position to ensure that your heels are touching the toe-touch line and that the measuring tapes are in a direct line with your body.

6. Without bending your knees, stretch your arms and body forward toward your toes, extending your fingertips to the greatest possible distance.
7. Designate a group member to place a ruler vertically from the tips of your outstretched fingers to the tape on the floor and place a pencil mark where the ruler hits the tape. Now read the distance shown on the tape to be +, 0, or − the actual number of centimeters required to reach the toe-touch line.
8. Repeat the toe-touch test three times; use the best distance in three trials to record on the group record sheet.
9. After all members of the group have performed the test and recorded their best performance, analyze the results and discuss with other group members:

   - possible reasons for differences in performance on the three trials for the toe-touch test,
   - possible ways to increase flexibility ratings,
   - how increased flexibility reduces the risk of torn muscles and other injuries to the body,
   - how good flexibility contributes to good physical fitness.

8-10 PERFORMING A TEST FOR EXERCISE TOLERANCE

Motivation: Different students respond differently to the experience of participating in a strenuous game of tennis or basketball, or to that of competing in a running or swimming event. Some individuals tire very quickly and spend a long period of time recovering after only a short period of exercise. Others can exercise for long periods of time without becoming fatigued and require only a short time to recover from the stressful demands that have been made on their bodies. Performance on a test for rating exercise tolerance—the ability of the heart rate to slow down after stressful exercise—is a useful indicator of an individual's level of physical fitness.

Recommended Grade Level: Grades 7–8

Strategies Involved: Hands-on activity
Personal involvement
Student interaction
Science skills development

Materials Required:

Each student/partner combination will need:

- a low step stool or wooden platform 20 cm high
- a stopwatch or clock with a sweep second hand
- a copy of the STEP TEST RECORD AND EXERCISE TOLERANCE WORKSHEET
- a metronome (optional)

> NOTE: School health records should be consulted to identify students who, because of health problems, should not participate in the physical aspect of the step test. These students can be designated as the official record-keepers.

Procedure:

Instruct students, working with partners, to prepare for the test:

1. Work out and practice a perfectly-timed stepping rhythm with your partner who, using a stopwatch, will call out the proper timing to ensure that the steps you take will occur at precisely timed half-second intervals. A metronome may be used to help you establish the proper rhythm.
2. With your partner, practice the pulse-count technique for determining heart rate:

   - Rest your partner's wrist, palm of hand facing up, in the palm of your upturned hand.
   - Gently curl your fingers around the inside of his/her wrist so that your fingertips exert gentle pressure against the artery that is located in the area where his/her thumb joins his/her wrist.
   - Locate the point at which the beat is strongest and can be felt as a continuous activity.

STEP TEST RECORD AND EXERCISE TOLERANCE RATING WORKSHEET

NAME	PHASE 1		PHASE 2		PHASE 3		After 1-min. rest Pulse Rate/ 1 min.	Recovery Score	Exercise Tolerance Rating*
	Pulse Count/ 6 sec.	Pulse Rate/ 1 min.	Pulse Count/ 6 sec.	Pulse Rate/ 1 min.	Pulse Count/ 6 sec.	Pulse Rate/ 1 min.			

*Recovery score more than 10 = GOOD EXERCISE TOLERANCE RATING.
Recovery score of 10 or less = FAIR EXERCISE TOLERANCE RATING.
Inability to complete test = POOR EXERCISE TOLERANCE RATING.

- Count the number of pulse beats for exactly 6 seconds, as timed by a stopwatch or clock with a sweep second hand.
- Multiply this number by 10 and identify the resulting number as the normal pulse rate per minute for your partner.

3. Repeat steps 1 and 2, this time reversing the roles played by you and your partner.

Instruct students, working with partners, to perform the test:

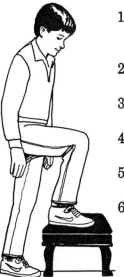

1. Stand directly in front of the step stool, with both feet flat on the floor and both hands held loosely at your sides.
2. On your partner's command UP, place your left foot on the stool.
3. On the count of TWO, bring your right foot up to assume a standing position on the step stool.
4. On the command DOWN, place your left foot on the floor, returning it to its starting position.
5. On the count of FOUR, return your right foot also to its starting position on the floor.
6. Follow the commands being given rhythmically every half second, completing the cycle: UP, TWO, DOWN, FOUR, every 2 seconds, and continuing the activity without any break in rate or rhythm for a period of 1 minute.

CAUTION: IF AT ANY TIME DURING THE TEST A FEELING OF DIZZINESS OR OVERTIREDNESS IS EXPERIENCED, STOP THE ACTIVITY AT ONCE AND ALLOW YOUR PARTNER TO HELP YOU TO A CHAIR WHERE YOU CAN SIT QUIETLY.

7. The moment you complete the 1-minute test, sit down and allow your partner to count your pulse beats during a 6-second period and to record this number appropriately for Phase 1 on the record sheet.
8. To continue the test, proceed to Phase 2: repeat the 1-minute step test, followed by a pulse count for 6 seconds, counted and recorded as before.
9. Repeat the procedure—1-minute step test, 6-second pulse count, and recording—one more time to complete Phase 3.
10. Then, sit quietly for 1 minute. At the end of this rest period, make a final determination of a 6-second pulse count and record the information appropriately.
11. In a similar manner, collect and record data for your partner's performance on the step test.

Instruct students, working with partners, to process the data on the worksheet:

1. For each 6-second pulse count taken, calculate and record the corresponding pulse rate per minute.
2. Calculate the RECOVERY SCORE for each participant by finding the difference between the pulse rate immediately following Phase 3 of the step test and the pulse rate after the 1-minute rest period.
3. Determine the EXERCISE TOLERANCE of each individual, as indicated by his/her RECOVERY SCORE.

As a follow-up, have all students share their experiences by participating in an open class discussion which focuses on:

- reasons for counting pulse beats for periods of 6 seconds rather than a full minute,
- factors that affect a person's tolerance for exercise,
- benefits to the individual who has a good exercise tolerance,
- how a good exercise tolerance rating can be used as an indicator of a person's physical fitness.

9

Individual
and Group Activities

A commitment to bringing about the involvement of all students in science requires a multifaceted, activity-oriented curriculum. It implies that the curriculum design should have a highly individualized approach that encourages students to deal with scientific issues to which they attach importance—thus maintaining the eagerness to find answers to questions demonstrated by most students when they are wide-eyed and curious youngsters in the primary grades, and, at the same time, awakening latent interests among those students who previously had been unmotivated or uninterested in science. Such a commitment implies also that provisions should be made for the involvement of *all* students in challenging activities through which they can gain valuable experience with science investigations. In this way they will gain some insight into how scientists think and work and they will come to realize that they, too, can use the scientific method in their approach to everyday problems.

It is important that students be involved in every aspect of an individualized activity. With proper guidance, they should make decisions relating to the selection of the study topic, and they should be actively involved in planning for its development and the manner in which the findings will be reported.

The topic for the activity should be chosen with care; it should capture the interest of the student and be relevant to his/her experience. Often, special interests are revealed during an open-ended class activity, during which a student's question can be turned into a challenging topic for him/her to investigate in greater depth. In addition, student hobbies, pastimes, reading preferences, and special talents may suggest a suitable topic for study. Students should be matched to an established interest and encouraged to develop it further by using a method or technique that he/she finds enjoyable and rewarding.

The most successful science activities are keyed directly to student interests, expressed by individuals or shared by classmates, either in small groups or as an entire class project. All students, however, should play a part in all activities, either as active participants or as interested spectators.

ACTIVITIES FOR THE ENTIRE CLASS

9–1 CONDUCTING A CONTEST BETWEEN BOYS AND GIRLS

Motivation: Several differences observed in the abilities of boys and girls can be traced to their basic physical structure and development. Boys, for example, can lift objects more easily because their longer bones provide greater leverage, and they can throw a ball a greater distance because of their greater muscular development. On the other hand, girls' handwriting is generally more legible, due to better coordination resulting from the earlier fusion of bones in their wrists. A contest that challenges boys and girls, propped at an angle against a wall, to regain an upright standing position without toppling, provides an opportunity to discover which sex has a body structure that offers a slight advantage for winning this competition.

Recommended Grade Level: Grades 7–8

Strategies Involved: Hands-on activity
Student involvement
Student interaction
Science skills development

Materials Required:

- a flat wall space in the classroom
- a student-prepared chart that provides for keeping score of successful performances for each team

Procedure:

Involve all members of the class in a contest that provides for competition between boys and girls:

1. Divide the class into two teams, boys vs. girls.
2. Conduct the contest so that two contestants, one from each team, attempt the feat at the same time. Instruct the two contestants to proceed as follows:

- Stand, facing the wall, with toes touching the wall at floor level.
- Step back from the wall a distance of 3 foot-lengths, measured by the contestant's own feet.
- Place both hands against the wall and lean forward until the top of your head touches the wall.

- Remove your hands from the wall and let your arms hang loosely in the area between your body and the wall, so that your body is now supported against the wall by your head.
- Without using your hands, bending your knees, or moving your feet, stand up straight.

3. Have one student, acting as class secretary, record on the chalkboard the ability or inability of each member of each team to perform successfully in the competition.
4. Continue, as in steps 2 and 3, until all members of the class have participated.
5. Instruct students to calculate the percentage of successes for the boys and for the girls, and have the class secretary record this data appropriately on the chalkboard.
6. As a follow-up, have all students analyze the data collected and participate in a class discussion in which they consider:

- which team scored better in the competition,
- the experiences reported by those who were unable to perform successfully,
- the differences noted in successful and unsuccessful performances, such as the angle between the body and the wall and the distance of the feet from the wall,
- the relationship between the scientific basis for this competition and the focus of the science topic involved in Activity 8–2.

9–2 PACKAGING FRAGILE OBJECTS

Motivation: Many fragile objects can be protected from breakage and destruction by placing them in specially designed packages. Cookies, for example, may reach their destination unbroken when carefully placed at precise distances from each other on a contoured plastic tray within their outer wrapper, and a flight recorder usually survives a plane crash when it is encased in the aircraft's "black box." Eggs, too, can be packaged in such a way that they are protected from breaking when their container is dropped. Designing such a package presents a stimulating challenge to students to exhibit creativity while applying their scientific knowledge to an original design.

Recommended Grade Level: Grades 6–8

Strategies Involved: Hands-on activity
                        Student involvement
                        Science skills development

Materials Required:

Each student will need:

- an uncooked egg
- a cardboard box (6 by 6 by 6 in. or larger)

- a crayon or colored pencil
- an available supply of cotton (balls or roll), aluminum foil, newspaper, styrofoam, polyethylene bubble pack, shredded paper, and carton tape
- access to a stepladder, about 2 meters high
- a tape measure

Procedure:

Instruct students to proceed as follows:

1. Design and construct a package that will protect a fragile object from damage due to a crash:

   - Examine your egg and cardboard box, and the assorted packing materials that are available.
   - On paper, draw a diagram of your design for safe packaging of your egg, indicating the materials used, as well as precise positions, thicknesses, and other specifications.
   - Obtain the materials needed from the available supplies.

   - Construct your package, as specified in your design.
   - With a crayon or other colored marker, draw an arrow on the outside of the box, indicating the "UP" position of the box by the point of the arrow.

2. Test your "crashproof" package by holding it in an upright position and dropping it from the top of a ladder* to the floor below.

3. Examine your package and its contents. Then, indicate on your diagram whether or not the design and construction of your package proved to be crashproof.

4. When time is provided, report to the class your findings:

   - factors contributing to a good design and construction,
   - improvements needed in the design and/or construction of a package that failed the test.

As a follow-up, engage all students in an open class discussion in which they focus attention on the procedure of designing, constructing, and testing packaging for the protection of breakable objects. They should consider:

- the relationship between the position of the objects and the protection provided,

*CAUTION: STUDENTS MUST BE CAREFULLY SUPERVISED WHEN USING A LADDER.

- the nature and amount of protective material used,
- the reinforcement of outside edges of a package,
- some actual designs and construction used in packaging eggs, cookies, fruit, and other fragile objects.

9–3 PREPARING AND DEPOSITING A SCIENCE TIME CAPSULE

Motivation: Excavations reveal much information about conditions that existed in past ages. Some articles buried in the pyramid tombs of Egyptian pharaohs, for example, give evidence of the level of scientific development that had been reached by that early civilization. Today it is a fairly common practice to fill the corner-stone of a large building with documents and objects that will, at some future date, be looked upon as artifacts of our society.

Many articles that are commonly used at home and at school may indeed become "artifacts" in a relatively short period of time. Preparing a science time capsule, with plans for digging it up on high school graduation day, enables students in the middle school to focus on the rapid rate at which our science and technology bring about changes in design and, in some cases, cause some objects to become obsolete.

Recommended Grade Level: Grades 4–8

Strategies Involved: Hands-on activity
Student involvement

Materials Required:

- a large metal container with a lid
- a paint brush
- rust inhibitor paint
- a liquid sealant and applicator
- a collection of assorted science-related objects, such as a science magazine summarizing the highlights of the current year, a science textbook and/or student notebook, a science videotape, some student science papers, and a currently popular scientific toy or puzzle
- aluminum foil
- boxes in assorted sizes

Procedure:

Appoint a committee of student volunteers to plan and conduct a class project that will involve all members of the class. Instruct the committee to proceed as follows:

1. With input from the class, select a location on school grounds and obtain permission from the school authorities for burying a science time capsule and marking the site and the official school records to identify the project.
2. Obtain a metal container that can be sealed and clean it thoroughly. Then, have the container painted, both inside and out, with a rust inhibitor paint.

3. Have an additional coat of paint applied to the outside of the container and its lid.
4. Have the class make a decision about what science-related objects to bury. In addition to a science book, a magazine, a current videotaped science program, a newspaper account of a recent scientific discovery, and selected pictures and descriptions of the latest developments in automobile and airplane design, be sure to include a student science notebook and some things to represent a favorite lab experiment or demonstration.
5. Have students wrap all items carefully, using aluminum foil and/or boxes, as needed, and place them carefully in the prepared container.
6. Have the lid placed on the container and have several coats of liquid sealer applied in order to seal it completely.
7. Hold a burial ceremony at the selected site and set a date, such as high school graduation day, when the class will meet to dig up the capsule, examine its contents, and assess the changes that have been brought about by scientific advancements during the intervening years.

9–4 GAINING AN ADVANTAGE BY USING A PULLEY SYSTEM

Motivation: The use of simple machines for making work easier is incorporated into some of the most common activities encountered in daily life. Both a stairway and a slanting road or pathway make climbing to the top easier. Similarly, pushing a heavy object up a ramp is easier than lifting it vertically, and using a lever to pry off a jar top makes this task easier to perform. In an activity that demonstrates the mechanical advantage of a pulley system, even the smallest girl in the class can overcome the combined physical force being exerted by two of her strongest classmates.

Recommended Grade Level: Grades 4–8

Strategies Involved: Hands-on activity
Student involvement
Student interaction
Science skills development

Materials Required:

- two broom handles with smooth edges
- a piece of sturdy rope, about 1 meter in length

Procedure:

A few days prior to the class activity, enlist the aid of three or four student volunteers for preparing a Tug-of-War device:

1. Attach one end of a rope to a broom handle.
2. Then, wrap the rope around a second broom handle.
3. Make two more turns of the rope, from one handle to the other, keeping the handles about 15 centimeters apart, and with one end of the rope remaining free and unattached.

On the day of the competition, select three students to participate actively in a Tug-of-War contest:

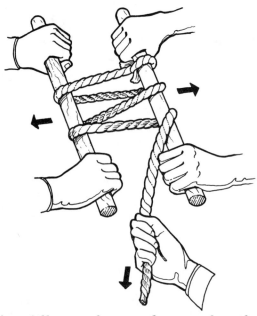

1. Two students, each holding a broom handle and pulling in opposite directions, try to separate the handles.
2. At the same time a third student, gripping the free end of the rope, pulls on the rope in an attempt to draw the broom handles together.
3. When the broom handles have been separated, or drawn together, a winner in the Tug-of-War contest is declared.
4. The Tug-of-War is repeated many times, allowing for different students to compete in the two-against-one competition.

As a follow-up, have students analyze the results of numerous contests and engage in an open class discussion to consider:

- the consistency of the results obtained,
- the identification of the simple machine incorporated into the design of the Tug-of-War device,
- how the mechanical advantage of this built-in pulley system could be increased or decreased by changing the number of turns of the rope around the broom handle.

9–5   BALANCING AN EGG ON END

Motivation: Balancing an egg on end so that it stands in an upright position without toppling is a delicate operation to perform. Christopher Columbus is said to have accomplished this feat over 400 years ago, astonishing his contemporaries, who thought it impossible to do. Today students face the challenge with varying degrees of success at the time of the spring equinox when, it is claimed, the earth's perfect balance—day and night being of exactly equal length—makes possible the balancing of an egg on end as well.

Recommended Grade Level: Grades 4–8

Strategies Involved: Hands-on activity
                            Student involvement
                            Investigative approach
                            Science skills development

Materials Required:

Each group of four students will need:

- four uncooked eggs (one per student)
- a flat surface, such as a table top

Procedure:

Instruct students, working in groups of four, to proceed as follows:

1. Check a calendar or an almanac to determine the exact date and time of the spring equinox.
2. Bring from home uncooked hens' eggs, sufficient in number to provide one egg for each member of your group to perform the test*:

   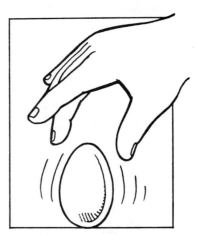

   - Examine your egg and locate its large end.
   - Hold the egg gently, with its large end resting on a table top that has been cleared of all books and papers.
   - Exercising care and patience, attempt to balance your egg on end so that it will remain in an upright position when you remove the hand with which you have been supporting it.

3. After several trials, compare your results with those of other group members at your table.
4. Discuss the group findings and consider:

   - if the belief that an egg will balance on end at the time of the spring equinox appears to be supported by the actual performance results,
   - how an eggshell that is slightly bumpy might influence the results,
   - what advantage might be gained, using Columbus' technique of shaking the egg before attempting to balance it on end,
   - if the "balancing" results are the same during other times of the year,
   - if the activity engaged in represents a scientific approach to problem solving.

"MAKE AND TAKE" ACTIVITIES

9–6 PROPAGATING PLANTS FROM CUTTINGS

Motivation: Most plants are grown from a tiny seed which, when planted in soil and supplied with suitable conditions of light, air, moisture, and temperature, will

---

*Depending on the exact time of the spring equinox for different locations and in different years, it may be necessary to perform this part of the activity at home.

develop into a young seedling and eventually form a plant like the parent stock that produced the seed. A few species, like the coleus, jade, and African violet, can also produce an entire new plant from a small section of the original. When coleus cuttings from the parent plant are placed in water, a new root system for the leaf and stem portion will develop. Similarly, African violet or jade leaves can be induced to develop roots from which tiny leaflets spring upward to form a tiny new plant that is exactly like the original. Spider plants, too, have devised ways to reproduce without seeds. Long, arching stems that blossom at their tips are sent out and, when these blossoms mature, they give rise to leaf and root systems at the end of each tip. If separated from the parent plant and provided with suitable conditions, each young plantlet will begin to grow and engage in life activities on its own.

Recommended Grade Level: Grades 4–8

Strategies Involved: Hands-on activity
Student involvement

Materials Required:

- several mature spider plants, which will provide one plantlet per student
- 4-inch flowerpots (one per student)
- potting soil or garden soil
- trowels or other gardeners' implements
- water
- "plant" food preparation

Procedure:

Instruct students to proceed as follows:

1. Working in small groups:

   - Place a mature spider plant, with plantlets, in the center of a table.
   - At each point where an arching stem from the plant has formed a plantlet, place a flowerpot that has been partially filled with soil.

   - Carefully guide each plantlet into its designated flowerpot and add additional soil to anchor the plantlet.
   - Water each plantlet in its new "home."
   - Allow the plantlets to adjust to their new homes.
   - Then, cut the umbilical cords and set each of the baby plants in an area that provides suitable conditions for its growth.
   - Care for the young plants, providing proper conditions of light, temperature, water, and "plant food," as needed.

2. Discuss with other members of the class who have engaged in this activity:

   - how the baby plants resemble the parent,
   - how the time required for producing a new plant by this method compares with that involved when a new plant is grown from a seed.

3. Distribute the plants so that each member of the class can take one home to share with family and friends.

9-7   WATCHING BIRDS AT A FEEDING STATION

Motivation: Interest in bird-watching is widespread. Organized field trips to wildlife refuges and bird sanctuaries are well-attended, and the annual "Big Day" event sponsored by bird clubs is popular among club members, many of whom tabulate the number of different species spotted during the designated 24-hour period. Most bird-watching, however, takes place more informally in gardens and backyards, where songbirds can be attracted by supplies of suet and seeds displayed at conveniently placed bird-feeding stations.

Recommended Grade Level: Grades 4–6

Strategies Involved: Hands-on activity
Student involvement

Materials Required:

Each student engaging in the activity will need:

- a clean, empty mesh or string bag, such as an onion bag from the produce department of the grocery store
- several pieces of suet from the meat department of a food store
- a nylon thread or fishing line, about 1 meter in length
- a Field Guide to Birds, or other appropriate bird identification guide
- a meter stick
- a notebook for recording observations made

Procedure:

Instruct students to proceed as follows:

1. Place several pieces of suet in the open end of a clean mesh bag, such as an onion bag.
2. Draw the bag closed at the top, or gather the edge of the bag at its open end and wrap a piece of thread around it.
3. Tie a strong cord or nylon thread around the top of the bag, leaving an additional length of 60 to 75 cm of cord to serve as a hanger.
4. Tie the free end of the cord to a tree branch or other structure where it can be viewed from a window in the classroom or at home.
5. Observe the kinds of birds that visit the "feeding station":

    - Note the size, color, and other features of the visiting birds.
    - Use reference books to identify the birds viewed.
    - Keep a record of the names of the birds and of the dates they were observed.

6. Maintain the "feeding station" throughout the season, replacing the suet, as needed.

7. On the specified date, bring the record of birds observed to class and discuss with other class members:

   - the number and variety of birds observed,
   - any changes noted in the kinds of birds visiting the feeding station at the beginning of the season and at the end,
   - the importance of continuing the food service once this practice has been initiated.

9–8 CREATING "SNOW" BY ALLOWING SOME PARTICLES OF A SOLID IN A LIQUID TO SETTLE

Motivation: Often it is difficult to detect the presence of a colorless solid substance that has been dissolved in water. When particles of the two materials become intermingled, the solid seems to disappear as its molecules become evenly distributed throughout the water. There is, of course, a limit to the amount of solid that can be dissolved in a given amount of water at a given temperature. After the saturation point has been reached, some interesting effects produced by any excess solid particles that are present can be observed.

Recommended Grade Level: Grades 4–6

Strategies Involved: Hands-on activity
Student involvement
Science skills development
Student interaction

Materials Required:

Each student engaging in the activity will need:

- a small glass jar (about 8 ounces) with a screw-cap cover
- boric acid crystals
- water
- food coloring
- a small figurine
- a teaspoon measuring spoon
- waterproof glue

Procedure:

Instruct students to proceed as follows:

1. Select a suitable figurine that will fit inside a small glass jar and, using waterproof glue, attach the figurine to the inside bottom of the jar.
2. Allow the glue to dry so that the figurine remains firmly attached to the floor of the jar.

3. Place 5 teaspoons of boric acid crystals in the jar.
4. Pour water into the jar until it is completely filled.
5. Add one drop of food coloring.
6. Place the top on the jar and screw it tightly in place.
7. Shake the jar vigorously.
8. Observe the contents of the jar as the "snowflakes" settle.
9. Take the snowscene home for showing or for gift giving, with an explanation of:

- why some boric acid crystals float around and then sink after the jar has been shaken.
- why the solution of boric acid is called "saturated."

9–9   GROWING A HANGING SPONGE GARDEN

Motivation: Plants are traditionally grown in soil containing nitrogen, phosphorus, potassium, and other chemical substances needed for their proper growth and development. It is not unusual, however, to find plants being grown in a soilless culture; ornamental plants, such as paperwhite narcissus, can be grown from bulbs placed in undrained bowls filled with pebbles, and greenhouse plants of many species are grown in nutrient solutions. The ever-diminishing amount of land available for traditional farming methods portends a future dependence on the use of similar hydroponic methods for large-scale growth of agricultural crops as well.

Recommended Grade Level: Grades 4–6

Strategies Involved: Hands-on activity
Student involvement
Student interaction
Science skills development

Materials Required:

Each student engaging in the activity will need:

- a natural sponge whose shape resembles a ball
- a 2-foot length of nylon thread or fishing line
- a large needle
- a pan or bucket of water
- a large colored bead
- birdseed mixture

Procedure:

Instruct students to proceed as follows:

1. Obtain a natural sponge whose shape resembles a ball.
2. Make a knot at one end of a length of nylon thread or fishing line and slip a large bead next to the knot.
3. Now thread the line through the center of the sponge, from one side to the other.
4. While holding the sponge above an area spread with newspaper, shake birdseed over the sponge and collect any seed that does not attach to the sponge.
5. Gently, dip the sponge in a pan or bucket of water, being careful not to dislodge any of the seeds.
6. Hold the sponge over the pan or bucket, allowing the excess water to drip back into the vessel.
7. Use the nylon thread to hang the sponge from a hook or other support near a sunny window.
8. Each day hold a pan of water under the sponge and allow it to absorb water, as needed.
9. Examine the sponge each day and look for small plantlets to develop as the seeds sprout.
10. When your "sponge garden" has become established, take it home and hang it near a sunny window where it can be observed and cared for daily, following the same pattern used in the classroom.
11. Report interesting developments as your sponge garden grows and compare these developments with those being reported by other students who are also growing sponge gardens.

9–10 "GROWING" A CHEMICAL CRYSTAL GARDEN

Motivation: Although only living things engage in true growth, under certain conditions some chemical substances also appear to grow. For example, when a supersaturated solution of a substance either cools or loses some of its liquid by evaporation, the solid particles settle out in an orderly pattern and become consolidated. Particle by particle the substance is built into its characteristic crystalline form—table salt in small cubes, quartz in column formation, and magnesium in a feathery pattern. Although a microscope must be used to witness the actual building process of crystals, no magnification is needed to view the overall effects of their formation in the growth of an attractive crystal garden.

Recommended Grade Level: Grades 5–8

Strategies Involved: Hands-on activity
Student involvement
Student interaction
Science skills development

Materials Required:

Each student engaging in the activity will need:

- four or five charcoal briquettes
- household ammonia
- table salt
- laundry bluing solution
- food coloring
- water
- a shallow pan
- a cup
- a tablespoon for measuring and mixing

Procedure:

Instruct students to proceed as follows:

1. Place four or five charcoal briquettes in the bottom of a shallow pan.
2. In a cup, mix together:

   1 tablespoon of household ammonia
   2 tablespoons of water
   1 tablespoon of table salt
   2 tablespoons of laundry bluing solution.

3. Pour the prepared mixture over the charcoal briquettes.
4. Sprinkle a few drops of food coloring over the surface of the moistened briquettes.
5. Set the pan in an area where it can remain undisturbed overnight.
6. Next day, examine the pan for evidence of change:

   - signs of evaporation of some of the liquid,
   - appearance of crystals beginning to form on top of the charcoal.

7. Continue to examine the pan on a daily basis and, over the next few days, compare your "crystal garden" with those prepared by other students.
8. Discuss with other members of the class the changing appearance of the "gardens" and the process by which they "grow":

- the size and shape of crystals formed,
- the effect of water evaporation on the formation of crystals,
- reasons for different colors of crystals,
- examples of crystal formation in other situations, both natural and man-made,
- differences between "growth" of crystals and growth of a plant or animal.

9. Take your crystal garden home and display it for your family and friends to enjoy as it continues to "grow."

ACTIVITIES FOR INDIVIDUAL STUDENTS

9–11   PERFORMING A SHOW-AND-TELL DEMONSTRATION

Motivation: Every student in the middle grades has acquired some scientific knowledge that can be shared with others. Demonstrating how to fly a kite or use a skateboard, explaining the dynamics of the flight pattern of a Frisbee, and describing the structural features used to arrange a collection of rocks, butterflies, or seashells into related groups are examples of activities that students can report on from within their personal experience. Presented as a variation of the "show-and-tell" approach youngsters find so exciting in the primary grades, these demonstrations offer a change-of-pace type of activity in which students learn from each other.

Recommended Grade Level: Grades 4–8

Strategies Involved: Student involvement
Student interaction
Independent study approach
Science skills development

Materials Required:

- an assortment of objects found on the Science Display Table or brought in by students
- copies of SUGGESTED TOPICS FOR A SHOW-AND-TELL DEMONSTRATION (one per student)
- classroom and library reference materials

Procedure:

Instruct students to proceed as follows:

1. Select an object in which you have an interest or about which you are curious.
2. Investigate the object and learn as much as you can about it.
3. Prepare a presentation in which you plan to show the object to the class and tell them about it.

4. Deliver your SHOW-AND-TELL presentation:

- Show your object and tell what it is.
- Demonstrate and explain how it works.
- Tell how it is related to science.
- Tell how it can be used.
- Answer any questions other students may ask about it.

5. Display your object where all students can see and examine it.

**SUGGESTED TOPICS
FOR A SHOW-AND-TELL DEMONSTRATION**

A periscope

A gyroscope

A music box

Magnetic jumping disks

Mexican jumping beans

A hologram paperweight or magazine cover

A kaleidoscope

A yo-yo

A bubble hoop

A floating magnet

A fiber optics rod

A paper airplane

A Frisbee

A "dunking bird"

A collection of rocks, minerals, insects, seashells,
    fossils, or other science-related objects

9–12 PREPARING A SCIENCE PROJECT

Motivation: Science projects are the most popular form of student involvement in individualized activity. They can be initiated on an individual or small group basis at any time during the school year, as suggested by a topic under study, an unusual observation or occurrence, or a problem encountered during the events of the day. Students should be encouraged to pursue the topic of concern, consulting available reading and reference material for information relating to the topic and using the knowledge gained as a base upon which to build their project study.

When placed on display during an OPEN HOUSE or SCIENCE FAIR exhibit, science projects represent in graphic form a wide array of interests, developed by students in a variety of ways. Although it is common practice to judge the merit of the project displays, recognition should be given to all student efforts. Rather than declaring "winners," the awarding of ribbons and/or certificates of accomplishment to all who participate is more effective for motivating students to become independent learners. By learning to think and act like scientists, they become adept at using scientific approaches in their search for answers to questions and solutions to problems in a variety of life situations.

Recommended Grade Level: Grades 4–8

Strategies Involved: Hands-on activity
Student involvement
Independent study approach
Science skills development

Materials Required:

- copies of GUIDELINES FOR PREPARING A SUCCESSFUL SCIENCE PROJECT (one per student)
- classroom and library reference materials
- assorted materials for student assistance in conducting the project study and preparing an exhibit
- a supply of ribbons and certificates to be awarded to student participants
- notebooks or folders (one per student)

Procedure:

Instruct students to proceed as follows:

1. Select for study a topic that is of particular interest to you.
2. Follow the GUIDELINES FOR PREPARING A SUCCESSFUL SCIENCE PROJECT.
3. Maintain a notebook or folder for recording all information related to the project.
4. Design and construct an exhibit that tells a story about your project.
5. Display your finished exhibit and be prepared to give an oral explanation of your project and to answer any questions that might be asked about it.

GUIDELINES FOR PREPARING A SUCCESSFUL SCIENCE PROJECT

1. DECIDE on a general topic area that interests you.

2. GATHER information about several aspects of the topic area.

3. SELECT one specific aspect of the topic to be the focus of your project.

4. PLAN your investigation, using your own ideas whenever possible.

5. BUDGET your time so that your project will be ready on time, with no need to rush at the last minute.

6. RECORD all information accurately about what was done and what happened.

7. ORGANIZE your materials in the form of a display that tells a story and is easily understood.

8. CONSTRUCT a freestanding background on which to mount your display materials.

9. POSITION your project display in the designated space and be prepared to give an explanation or to answer questions that may be asked by viewers.

10. REMEMBER that the purpose of your project is to prepare a display that viewers will find ATTRACTIVE, INTERESTING, and INFORMATIVE.

SUGGESTED TOPICS FOR A SCIENCE PROJECT

Metamorphosis of a Butterfly

Houseplants That Fight Air Pollution

Records in Fossils

Dyes from Unused Portions of Fruits and Vegetables

Raising Brine Shrimp

Protective Coloration in Nature

Crystal Formation

Bird Migration

Growing a Sweet Potato Vine

Investigating the Formation of Large Hailstones

Effects of Rock Music on Plant Growth

Making a Lemon Battery

Racing Cockroaches

Soilless Gardening

10

Learning to Live
and Work in the World
of Science and Technology

The world in which today's middle school students will be living their adult lives will be far different from any we have yet experienced. Therefore, we cannot know, or teach today's students, precisely what the world of the future will be like—which science discoveries will be made, which technologies will be developed, or which careers to plan and prepare for. Rather, we must prepare students for *change,* by concentrating on learning activities in which they will:

1. learn how to gain an understanding of new scientific knowledge and its implications, using a variety of available resource materials and methods of learning that are meaningful in their lives,
2. develop an awareness of the current trend from an Age of Technology to an Age of Information, and of the demand for inventiveness, creativity, and problem-solving abilities for living and working in a world in transition,
3. develop favorable attitudes so that they recognize the need for using an open-minded approach, are capable of accepting the inevitability of change, and will assume a responsible manner in the making of informed decisions, including those that concern man's stewardship of the earth and all of its resources.

ACTIVITIES FOR LEARNING HOW TO LEARN

10–1 MONITORING ONE'S OWN DENTAL HEALTH

Motivation: Technological advances in personal health care products and practices make it possible for individuals to monitor various aspects of their own state of health and physical fitness. A whole new industry has sprung up devoted to producing products with which consumers themselves are able to conduct tests that were formerly done only by professionals. Easy-to-use devices for determining lung capacity, grip strength, and blood pressure, as well as home test kits for determining early warning signs of some kinds of cancer, are among the many such consumer goods that are currently in use, and there are indications that the list will be extended to include many other health concerns as well.

Evaluating the effectiveness of techniques used to promote good oral hygiene can be demonstrated quite dramatically when students engage in an activity that detects the presence of plaque, which is generally associated with tooth decay.

Recommended Grade Level: Grades 4–8

Strategies Involved: Hands-on activity
Student involvement
Student interaction
Science skills development

Materials Required:

Each group of four students will need:

- plaque-indicator pills (one per student)
- personal toothbrushes (one per student, brought from home)
- small paper cups (one per student)
- personal tubes of toothpaste (one per student, brought from home)
- a package of dental floss
- a box of toothpicks
- water
- a sink or basin
- a mirror

Procedure:

Instruct students, working in groups of four, to proceed as follows:

1. Working individually, test the effectiveness of various methods of cleaning the teeth:

   - Chew a plaque-indicator pill thoroughly.
   - Rinse your mouth with water and observe how much of the red coloring matter remains on the teeth.

- Brush your teeth with a back-and-forth motion, using a personal toothbrush and the brand of toothpaste normally used at home. Then, rinse the mouth again and, using a mirror, observe the amount of red coloring still remaining on the teeth.
- Brush the teeth again, this time using an up-and-down stroke. Then, again rinse the mouth and observe the remaining red color.
- Use a toothpick to clean places in the mouth not cleaned by the toothbrush. Again, rinse with water and observe for traces of red color.
- Use dental floss between the teeth and again rinse with water and check for traces of color.

2. On the basis of the test results, make a personal evaluation of the methods used for cleaning the teeth:

- Determine if any one method removed all of the red coloring.
- Determine whether some methods of cleaning appear to be more effective than others.
- Plan a daily routine to be used personally for thoroughly cleaning the teeth between visits to the dentist.

3. Discuss with other members of your group:

- how the red color indicates that the teeth are not thoroughly clean,
- how traces of food left on tooth surfaces and between the teeth provide a feeding ground for bacteria that cause tooth decay and bad breath,
- how individuals are able to assume greater responsibility for monitoring their own dental health by using plaque-indicator pills and other available consumer products.

4. On the basis of all information related to the activity, write a summary statement about what plaque is and how methods for its detection and removal can be used by individuals to monitor their own dental health.

10–2 READING LABELS TO MAKE INFORMED DECISIONS WHEN CHOOSING FOODS

Motivation: More and more, young people are being reminded to take care of their health while they are young and not to wait until they are 50. A sensible choice of foods in their early years might very well prevent difficulties later in life when they reach the age at which many people are faced with problems associated with cholesterol buildup, vitamin deficiencies, and/or an overweight condition.

In an effort to maintain a slim physique, currently in vogue, some people place great emphasis on counting the number of calories they consume each day. But this is not enough. All individuals should develop an awareness of the nutritional values in various foods as well as their calorie content. Because of FDA rulings, much of this information, together with the percentage per serving of U.S. recommended daily allowances of the stated vitamins and minerals, is available on product labels.

Recommended Grade Level: Grades 6–8

Strategies Involved: Hands-on activity
Student involvement
Student interaction
Science skills development
Interdisciplinary approach

Materials Required:

Each group of four students will need:

- nutrition labels from four different brands of cereal
- a copy of the chart PERCENTAGE OF THE U.S. RECOMMENDED DAILY ALLOWANCE OF VITAMINS AND MINERALS PER ONE-OUNCE SERVING OF CEREAL

Procedure:

Instruct students, working in groups of four, to proceed as follows:

1. Remove the nutrition labels from four different brands of cereal and mark each with a code (A, B, C, or D) that will identify it on the comparison chart.
2. Distribute the nutrition labels so that each member of the group has a different product for analysis:

   - Read the label from the cereal box.
   - Transfer necessary information from the label to the comparison chart.

3. Analyze the tabulated information and discuss with members of the group:

   - Based on a 1-ounce serving, which cereal comes closest to supplying the recommended daily allowance of the vitamins and minerals listed, as determined by the U.S. RDA?
   - How many servings of each cereal would be needed to supply all of the RDA of each vitamin and each mineral?
   - Would it be realistic (or even desirable) to expect a cereal to supply all of the RDA of each vitamin and each mineral?
   - What other foods are relied on to complement breakfast cereals as important sources of some of the vitamins and minerals needed each day?

4. Write a short paragraph in which you explain how nutrition labels on foods contribute to the ability of individuals to plan meals that will adequately supply their nutritional needs.

Name _____ Date _____

PERCENTAGE OF THE U.S. RECOMMENDED DAILY ALLOWANCE
OF VITAMINS AND MINERALS PER ONE-OUNCE SERVING OF CEREAL

Code	Name of Cereal	Vitamin				Mineral			
		A	B	C	D	Ca	Fe	Zn	P
A									
B									
C									
D									

10–3   GAINING SCIENTIFIC KNOWLEDGE TO IMPROVE ATHLETIC SKILLS

Motivation: Each Olympic year new records are set for sports events as competing athletes continue to display an ability to run faster, jump higher, and perform gymnastic events with greater precision than any athlete before them has ever exhibited. Advances in training techniques and related technological progress are largely responsible for these record-setting, championship performances.

Because of the ever-changing nature of technology and the ever-widening applications of scientific knowledge to sports science, which seeks ways of providing athletes with a slight edge, the future promises new record-breaking events in the Olympic games for many years to come.

Recommended Grade Level: Grades 7–8

Strategies Involved: Hands-on activity
Student involvement
Science skills development

Materials Required:

Each student engaging in the activity will need:

- a strip of paper that is 15 cm long and 2.5 cm wide
- classroom and library reference materials
- a metric ruler

Procedure:

Instruct students to proceed as follows:

1. Working individually,

- Prepare a strip of paper that is 15 cm long and 2.5 cm wide.
- Fold down a 2.5-cm flap at one end of the paper strip.
- Hold the flap against your lower lip so that the remaining part of the paper strip extends outward in front of you.
- Blow your breath gently, but steadily, while holding the paper strip firmly in this position.
- Note what happens to the paper strip.
- Analyze your findings:

   1) Does the air produced by blowing move faster over the upper or lower surface of the paper strip?
   2) How does this cause the paper strip to rise?

2. Discuss with other class members how this scientific knowledge can be used to advantage:

   • value in lifting an airplane off the ground during takeoff,
   • value to a ski jumper who positions his body in a slight curve over his skis so that the airflow passes over and under his body.

3. After consulting appropriate classroom and library reference materials, write a short paragraph explaining how scientific information, such as a knowledge of streamlining and leverage, is useful to a skier, skater, or other athlete.

10-4 DETERMINING THE ACTUAL FAT CONTENT IN A SERVING OF FOOD

Motivation: Interest in health and nutrition has increased to the point that most Americans are concerned with the potential dangers of high fat-intake and its relation to cholesterol buildup. Fast food restaurant specialties and packaged snack foods particularly have come under scrutiny for their high fat content, "empty" calories, and questionable nutritional value. Similarly, advertising claims need to be checked and the information contained on food product labels analyzed to determine the actual percentage of calories supplied by the fat portion of ingredients in a "lite" food product.

Recommended Grade Level: Grades 7–8

Strategies Involved: Hands-on activity
Student involvement
Student interaction
Interdisciplinary approach
Science skills development

Materials Required:

Each group of four students will need:

   • labels from both regular and "lite" forms of four different food products, such as soup mix, waffles, cheese, and salad dressing
   • four copies of the student worksheet CALORIES PROVIDED BY THE FAT CONTENT OF A SINGLE SERVING OF FOOD (one per student)

Procedure:

Instruct students, working in groups of four, to proceed as follows:

1. Obtain labels from both regular and "lite" forms of at least four different food products, such as soup mix, waffles, cheese, and salad dressing.
2. Read the information included in the listing of product ingredients on each label.

CALORIES PROVIDED BY THE FAT CONTENT
OF A SINGLE SERVING OF FOOD

Name of Product	Form	Number of Calories per Serving	Number of Grams of Fat per Serving	Number of Fat Calories per Serving	Percentage of Fat Calories
Ex: Whole Grain Waffles	Frozen	180	7	$7 \times 9 = 63$	$\frac{63}{180} = 35\%$
Buttermilk Waffles	Frozen	200	6		
1.					
2.					
3.					
4.					
5.					
6.					
7.					
8.					
9.					

3. For each form of each food product, locate and record in the proper spaces on the worksheet information concerning the number of calories and the grams of fat contained in a single serving of each food being investigated.

4. For each product listed on the worksheet, calculate the number of fat calories per serving and the percentage of calories that are provided by the fat content. Then, record this information appropriately to complete the worksheet.

REMEMBER: THERE ARE 9 CALORIES PER GRAM OF FAT.

5. Analyze the information on the worksheet and compare the percentage of fat in a serving of regular food with the percentage of fat in a serving of a "lite" formula for the same type of food.

6. Discuss the group findings with members of other groups. Consider:

- the importance of adhering to the guideline set by most nutrition experts, "No more than 30 percent of your total daily calories should come from fat,"
- whether the percentage of calories provided by fat in each product studied lies within this guideline,
- if there is a significant difference in the fat content of a regular and a "lite" form of food,
- how a product label might be misleading if the information it contains is not read with understanding,
- the significance of the percentage of fat in the daily diet,
- how reading health messages on food labels can help people avoid some health hazards.

10–5 CHECKING FOODS FOR TRACES OF CHEMICAL PRESERVATIVES

Motivation: Not all foods are "all natural." In addition to the vitamins, minerals, and nutritional supplements that are added to many foodstuffs, some chemicals may be used to improve the color and flavor and to prevent food spoilage due to decomposition. Traces of sulfur dioxide usually can be found on dried fruits that have been treated with this chemical preservative. Its presence can be detected by a simple chemical test.

Recommended Grade Level: Grades 7–8

Strategies Involved: Hands-on activity
                        Students involvement
                        Student interaction
                        Investigative/discovery approach
                        Science skills development

Materials Required:

Each group of four students will need:

- four or five pieces of dried fruit and information from the package label
- a 3 percent solution of hydrogen peroxide
- a 3M solution of barium chloride in a dropper bottle. (Prepare by dissolving 62.5 g of barium chloride in 100 ml of water and dispensing in small dropper bottles.)
- a tongue depressor
- distilled water
- two clean mayonnaise jars
- a graduated cylinder
- a coffee filter or several thicknesses of cheesecloth
- four copies of the Flowchart and Data Record sheet (one per student)
- access to a master Flowchart/Data Record form on the chalkboard

Procedure:

Instruct students, working in groups of four, to proceed as follows:

1. Early in the day:

   - Place four or five pieces of dried fruit in a clean glass jar. (Write the name of your fruit in the proper column on your Flowchart and Data Record sheet.)
   - Add distilled water to cover the fruit pieces and fill the jar to about the two-thirds full level. (Place a check mark in the proper space on your record sheet to show that the water has been added.)
   - Set the jar aside and allow it to remain undisturbed for about 4 hours, or overnight.

2. Later in the day, or the next day:

   - Examine the fruit pieces in the jar, noting any changes in the appearance of the fruit pieces and/or the amount of water in the jar.
   - Prepare a filtration system: support a coffee filter or several thicknesses of cheesecloth in the mouth of a clean glass jar.
   - Use a tongue depressor to press as much liquid as possible from the fruit pulp, and pour the liquid through the filter. (Place a check mark in the proper column on your worksheet to show that you have obtained a filtrate.)

Name _____   Date _____

ADDITIVES IN DRIED FRUITS
FLOWCHART AND DATA RECORD

Fruit	Step 1 H₂O → Filtrate	Filtrate +	Step 2 Hydrogen Peroxide +	Barium Chloride →	Appearance of Complex	Results Interpretation	Label Confirmation
Ex. Dried Apricots	✓	✓ → ✓	✓	✓	White Precipitate	Sulfur Dioxide Present	Confirmed
1.		→					
2.		→					
3.		→					
4.		→					

- Add 50 ml of 3 percent solution of hydrogen peroxide to the collected filtrate, and mix well. (Place a check mark on the second line of the filtrate column and one in the hydrogen peroxide column to show that this has been done.)
- Add, drop by drop, barium chloride solution and observe closely to determine if a white precipitate forms, indicating the presence of sulfur dioxide in the filtrate. Then, record your findings appropriately on your record sheet.

> CAUTION: REMEMBER TO HANDLE ALL CHEMICALS WITH
> CARE AND TO REPORT IMMEDIATELY ANY
> ACCIDENT THAT OCCURS.

3. Following the testing procedure, check the product label to confirm the results of the findings. (Write either Confirmed or Not Confirmed in the appropriate column.) Then, list all information appropriately on the chalkboard Flowchart and Data Record for tabulating all data collected by all groups.

As a follow-up, engage all students in a class discussion based on the data collected:

- agreement or disagreement of the test results with the product label information,
- identification of other additives indicated on the food labels,
- reasons for the use of additives in foods,
- advantages and disadvantages associated with the use of food additives.

After the discussion, have students transfer all information from the chalkboard Flowchart and Data Record onto their individual copies of the record sheet, which should then be placed in their science notebooks.

ACTIVITIES FOR DEVELOPING AN AWARENESS OF CHANGES BROUGHT ABOUT BY TECHNOLOGICAL PROGRESS

10-6 RECOGNIZING THE ADVANTAGEOUS FEATURES OF HYBRID ORGANISMS

Motivation: Many animals have characteristic features: the wool of the sheep, sharp teeth of the tiger, long legs of the horse, and trunk of the elephant are well-known to all. The interbreeding of two different species of animal has resulted, in some cases, in the best features of each being present in their hybrid offspring. For example, the mule exhibits the sure-footedness of its father (a donkey) and the strength and intelligence of its mother (a horse). The alpaca of South America has been crossed with the vicuna in order to produce a new animal, the paco-vicuna, which grows long, fine, woolly hair in greater quantities than the vicuna and of better quality than the alpaca. Recently, matings between a goat and a sheep, a tiger and a lion, and a whale and a dolphin have resulted in new hybrid offspring, named respectively the GEEP, TIGON, and WHOLPHIN. Undoubtedly, in the future, other matings of different species will produce additional new hybrids that exhibit a combination of the desirable features of each parent.

Recommended Grade Level: Grades 4–8

Strategies Involved: Hands-on activity
Student involvement
Student interaction
Science skills development

Materials Required:

Each group of four students will need:

- pictures of a variety of animals within a major classification
- four pairs of scissors (one per student)
- construction paper
- paste or glue

Procedure:

Instruct students, working in groups of four, to proceed as follows:

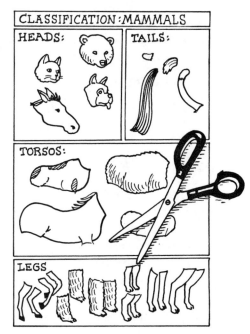

1. Select pictures of a variety of animals within a specific group, such as mammals, fish, reptiles, birds, insects, or some other classification.
2. With scissors, cut body parts of the animals and arrange them in separate stacks of heads, legs, beaks, wings, horns, or other body parts appropriate for your animal group.
3. On a piece of construction paper, reassemble a whole animal, designing it with structures and body parts that will offer it specific advantages. Then, glue the pieces in place.
4. Assuming that such an animal could be produced by genetic engineering, decide whether its introduction into the environment would be desirable or undesirable for man and for nature.
5. Designate one member of the group to present the new animal to the class and to describe the conditions under which it might be more successful than any related species presently known to man.

After all groups have presented their designs for "new" animals, form a committee of student volunteers to prepare a bulletin board display of the composite pictures.

10–7 INVESTIGATING SCIENCE-RELATED CAREERS

Motivation: Many students perceive a scientist as a man who wears a white coat and performs experiments in a laboratory. While only a small percentage of students will actually become scientists, an opportunity should be provided for all students to become aware of the wide variety of science-oriented and science-related careers currently open to both men and women.

As in the past, some of today's students express a rather high degree of certainty about "what they would like to be when they grow up." These career goals may or may not be realized; however, it is highly probable that an individual will not pursue the same kind of work throughout his or her working career. According to predictions made by the College Placement Council, college students will have, on the average, three different careers during a lifetime in the workplace—and this estimate may apply as well to those who do not attend college. New career goals may be set, and changes in technology may phase out many traditional kinds of work. This means that all students need to be adequately informed about a wide variety of careers and should prepare themselves to participate in general fields where the nature of the work is consistent with their special interests and abilities. Investigating science-based and science-related careers will provide students with a panoramic view of career choices.

Recommended Grade Level: Grades 4–8

Strategies Involved: Student involvement
                Student interaction
                Science skills development

Materials Required:

- copies of SCIENCE CAREER ANALYSIS WORKSHEET (one per student)
- information sources (interviews, library reference books, magazine articles and pictures, and other available sources)
- colored construction paper
- a bulletin board

Procedure:

Instruct students to proceed as follows:

1. Working individually:

   - Obtain a SCIENCE CAREER ANALYSIS WORKSHEET.
   - Use all available resources for gathering information needed, as indicated on the worksheet.
   - Complete the worksheet, including an illustration that shows a person actively engaged in some aspect of the career you have selected for investigation.
   - Mount your completed worksheet on a sheet of colored construction paper.

2. Working with members of a bulletin board committee, incorporate your career analysis with those of other students contributing to a bulletin board display on SCIENCE CAREERS.

As a follow-up, engage all students in an open class discussion of the advantages and disadvantages of careers investigated.

Name _____   **Date** _____

SCIENCE CAREER ANALYSIS WORKSHEET

1. The science activities I enjoy most allow me to work with

   _____

2. Some science careers in which these kinds of activities are involved include

   _____   _____

   _____   _____

3. One science career I think I would find interesting is

   _____

   because a person in that career field does the following:

   _____

   _____

   _____

4. I have interviewed a person in this career field, who reported: the most-liked features of
   this career:

   _____

   _____

   _____

   the least-liked features of this career:

   _____

   _____

   Illustration of a person actively
   engaged in this science career

ACTIVITIES FOR DEVELOPING RESPONSIBLE SCIENTIFIC ATTITUDES

10–8   CONTRIBUTING TO RECYCLING EFFORTS

Motivation: Recycling is fast becoming a way of life. Many communities have programs for recycling paper, glass, aluminum, and steel, and recently a new process has been developed for breaking down polystyrene foam, found in hamburger containers and insulated cups, to produce a plastic resin that can be used to form new plastic flowerpots and coat hangers. Many other "throw-away" products can also be used more than once. By participating in this kind of recycling, individuals can contribute to the overall effort to reduce the amount of energy being consumed as well as the amount of solid waste being produced.

Recommended Grade Level: Grades 4–8

Strategies Involved: Student involvement
Student interaction
Interdisciplinary/integrated learning approach
Reinforcement of learning

Materials Required:

- a collection of assorted articles slated for discard, such as onion string bags, empty milk cartons, broken shoelaces, dried-up ball point pens, empty juice cans, paper lunchbags, plastic food trays, margarine tubs, plastic bags, and non-returnable soda bottles

Procedure:

Instruct students to proceed as follows:

1. For a period of one week, bring in articles that have been slated for discard. Be sure that all are clean and safe to handle.
2. Examine all articles that have been brought in and displayed on the science table.
3. Select one item from the display and convert it into a usable product or suggest how it could be reused. For example, a clean, empty juice can might be used as a pencil holder, or a cut-down milk carton could be used as a planter. Then, add your idea for its reuse to the class list on the chalkboard.
4. When the class list is completed, make note of the many and varied uses for the items that had been slated for discard.

As a follow-up, engage all students in a class discussion that focuses on:

- additional uses for one type of item routinely discarded, such as plastic bags, making the list as long as possible by including suggestions made by all class members,
- how the reuse of materials that have been discarded contributes to the conservation of resources and to the solution of our problem of a shortage of landfill areas,
- how finding new uses for old items contributes to the development of a solid waste management plan, investigated in Activity 6-2.
- what individuals can do to give credence to the slogan "Recycling is Everyone's Concern."

10–9 INVESTIGATING METHODS OF SOLID WASTE MANAGEMENT FOR FOOD AND YARD WASTE

Motivation: The Environmental Protection Agency has issued some dire predictions about our nation's solid waste problem—while the amount of trash produced is steadily increasing, the number of landfills is declining sharply, and it is estimated that many states will have exhausted their available landfill space by the year 2000. Immediate action that involves everyone in programs of waste reduction, recycling, and resource recovery is needed to avert a crisis.

Recommended Grade Level: Grades 4–8

Strategies Involved: Hands-on activity
Student involvement
Student interaction
Interdisciplinary/integrated learning approach
Science skills development

Materials Required:

- assorted trash from one day's collection from home, the school lunchroom, the classroom, and the playground
- four large widemouthed jars (one per group)
- garden soil
- water
- labels for jars
- marking pens or pencils
- copies of the SOLID WASTE MANAGEMENT FOR FOOD AND YARD WASTE chart (one per group)

Procedure:

Engage all students in a class project:

1. Make a collection of clean trash from one day's accumulation obtained from home, the school lunchroom, the playground, and the classroom.

2. Assign each student to one of four landfill groups:

   a. Organic trash
   b. Reusable, recyclable trash
   c. Non-reusable, recyclable trash
   d. Non-reusable, non-recyclable trash

SOLID WASTE MANAGEMENT FOR FOOD AND YARD WASTE				
Percent (National)	**Category**	**Examples**	**Effects**	**Recommendation**
26	Organic	food scraps,	decompose	use for compost
20	Reusable, Recyclable	bottles, juice cans	do not decompose	recycle
46	Non-reusable, Recyclable	paper, cartons	decompose	convert to recycled paper products
8	Non-reusable, Non-recyclable	plastic rings, plastic rulers, plastic pens	do not decompose	reduce usage, find substitutes

3. Instruct each group to identify and claim articles of trash that represent the kind of trash specified for their landfill. Then, instruct each group to prepare a landfill, as assigned:

• Obtain a large, widemouthed jar.
• Select one or more articles of trash that will fit in the jar and place the article(s) inside the jar.

   • Add sufficient soil to cover the article(s).
   • Sprinkle water over the soil in the landfill.
   • Place a label on the jar to identify the type of landfill represented.
   • Place the landfill on the science table where it will be undisturbed, but accessible, for daily observations and additions of moisture to maintain its condition of dampness during a period of two weeks.
• Maintain a record of observations made during the two-week period.
• Appoint one group member to report to the class the findings concerning the type of landfill being investigated by the group.

4. Encourage students to consult available current literature and media resources, as well as bulletin board displays, that relate to methods of solid

waste management and the four landfill models being investigated in the class activity.

5. Then, encourage all students to participate in a brainstorming session in which they discuss:

- alternative methods for handling solid waste that could alleviate problems associated with landfill sites (which are fast becoming both unavailable and unwanted), depletion of natural resources, and rising costs,
- the NIMBY (NOT IN MY BACKYARD) syndrome,
- recommendations for effective handling of organic, reusable, and recyclable solid waste,
- responsibilities of individuals as well as those of government agencies.

10–10   USING A RENEWABLE ENERGY RESOURCE

Motivation: Among all the planets of our solar system, Earth is unique. For millions of years it alone has satisfied the environmental requirements for the support of life as we know it. Unfortunately, forecasts being made by some scientists indicate that the favorable conditions now provided by our planet may soon cease to exist. Environmental pollution threatens to upset the delicate balance of nature, and the supply of some widely used nonrenewable resources may soon be depleted. The responsibility of all of us to curtail activities detrimental to the environment and the urgent need to develop renewable energy resources are critical issues.

Recommended Grade Level: Grades 6–8

Strategies Involved: Hands-on activity
Student involvement
Student interaction
Science skills development

Materials Required:

Each group of four students will need:

- a bowl with a rounded bottom
- heavy aluminum foil
- a towel
- four wooden skewers or applicator sticks (one per student)
- marshmallows (one or more per student)
- a sunny location that provides direct sunlight

Procedure:

Instruct students, working in groups of four, to proceed as follows:

1. Working with other members of your group on a bright, sunny day:

- Line the inside of a bowl with aluminum foil and smooth the surface.
- Position the bowl in an area of direct sunlight so that the foil receives direct rays of light.

- Use a towel around the outside of the bowl to support the bowl in this position.
- Carefully move your hand from the inside center of the bowl outward, until you locate the position where the heat feels most intense.
- Place a marshmallow on the tip of a wooden stick applicator and, holding the stick end, maintain the position of the marshmallow where the heat appears to be most intense.

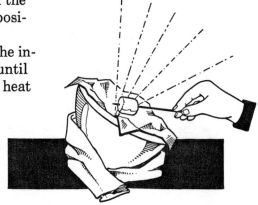

2. After all group members have toasted and enjoyed a marshmallow, compare the results experienced by various class members. Consider:

- why it was necessary to adjust the position of the bowl and support it with a towel,
- why it was necessary to focus the rays of sunlight,
- why it was necessary to reposition the bowl in order to toast the second marshmallow, . . . and the third, . . . and the fourth.

As a follow-up, engage all students in a class discussion, focusing on:

- the adaptability of the solar cooker to home cooking, and its limitations,
- other designs for solar ovens, solar panels, and solar cells,
- how solar energy is a beneficial source of energy that does not deplete our remaining fossil fuel resources,
- why solar energy is referred to as a renewable energy resource,
- how the development of renewable energy resources can help to avert an energy crisis in the future,
- how solar energy can play a part in solving our pollution problem,
- the advantages and disadvantages of solar energy as an alternative for fossil fuels.

10–11   ADOPTING A PLANT

Motivation: Caring for a pet, tending a small garden, and taking care of a house-plant are all rewarding experiences that have long-range, as well as immediate, benefits. In addition to gaining a certain satisfaction from observing the growth and behavior of a living organism, the sponsor learns to recognize special needs that must be met and characteristics that contribute to its success and well-being. In an expanded sense, the experience sets the stage for the individual to gradually assume the stewardship of the earth and all of its resources—both living and nonliving.

Recommended Grade Level: Grades 4–8

Strategies Involved: Student involvement
                                     Student interaction
                                     Student research and commitment

Materials Required:

- plant markers (one per student)
- copies of OFFICIAL PLANT ADOPTION CERTIFICATE (one per student)
- assortment of potted houseplants, such as Venus flytrap, geranium plant, sensitive plant, cactus plant, Boston fern, English ivy, and others
- reference books on care of houseplants

Procedure:

Instruct students to proceed as follows:

1. Observe the various plants that are displayed on the science table.
2. Select one plant that you would like to "adopt" and, by placing your name on a plant marker embedded in the soil of the appropriate flowerpot, obtain temporary custody of the plant for a period of one week, during which time no other student will be able to apply for custody of that plant.
3. Research information about your plant, including its proper care, light and soil preferences, watering needs, fertilizing needs, special growth characteristics, and susceptibility to disease and attack by enemies.
4. At the time specified for your appointment, make application to the class for permanent custody of the plant. Present your case by indicating:

   - the proper name of the plant,
   - the conditions known to be favorable for its best growth,
   - your willingness and ability to provide the favorable conditions and to avoid unfavorable conditions,
   - other convincing evidence that you will be a good guardian for the plant.

5. After receiving approval from the class, supply the necessary information to the class secretary, who will issue the official adoption papers.
6. Then, fulfill your obligations as guardian of the plant that has been placed in your care.

OFFICIAL PLANT ADOPTION CERTIFICATE

In the presence of a class member acting as your witness, indicate your intention to provide the necessary care and attention found to be in the best interest of maintaining the health and well-being of the plant you have chosen to adopt as your own.

This certifies that _____

(Name of plant)

was adopted by _____

(Name of guardian)

who is a member of _____

(Grade)

Signature of adopting guardian _____

Signature of witness _____

_____ Date

10–12 PREPARING TO ASSUME THE STEWARDSHIP OF THE EARTH

Motivation: The 20th century has been an exciting time, punctuated by technology that has produced television, videocassette recorders, computers, and supersonic aircraft, and that has paved the way for further progress associated with oceanographic explorations, space travel, and incredible advances in modern medicine. It is expected that the rate of technological change will accelerate at an even faster pace during the 21st century.

The students of today are the inventors and problem-solvers of tomorrow. Training and encouraging them to meet the challenges of a changing world will help them to prepare for the future when they will be called upon to consider alternative courses and to make responsible decisions.

Recommended Grade Level: Grades 4–8

Strategies Involved: Student involvement
Student interaction
Science skills development

Materials Required:

- a stack of Science Challenge cards

Procedure:

Instruct students, working in groups of four, to proceed as follows:

1. Select a Science Challenge card from the available stack of cards.
2. Read the science challenge presented to your group.
3. Brainstorm ideas relating to the problem presented:

   - Suggest related scientific information that might be helpful.
   - Consider how the model solution would be constructed and how it would work.
   - Consider possible flaws and weaknesses in the model and suggest ways to overcome them.
   - Prepare a diagram, a model, or a detailed description of your group's proposed solution to the challenge being met.

4. Select one group member to be the group spokesperson, who will present the group decisions to the class, explaining how the group's proposal might solve the problem posed and answering any questions that may be asked.

Following each group presentation, involve the entire class in an evaluation of the presentation to determine how well that particular group has met its challenge.

SUGGESTED TOPICS FOR SCIENCE CHALLENGE CARDS

Design a chalkboard that cleans itself.	Design an airplane that can take off on a runway that is only 50 feet long.	Design a kite that trails a message, announcing a special event such as the school science fair.
Design a robot that goes to the bookshelf and selects a book for you.	Design a bubble wand that makes multiple bubbles.	Design a device to separate tin cans from aluminum cans.
Design shoes that can be worn both in the classroom and on the football field.	Design a device that waters houseplants automatically while you are on vacation.	Design a device for preventing ice cubes from melting while they are being transported during warm weather.
Design a helmet that allows you to hear clearly while riding a motorbike.	Design a mechanical device for automatic feeding of goldfish.	Using only items usually discarded, build a cage suitable for a small animal.
Design a school bus that will fold, taking up only six feet of parking space in the school parking lot.	Design gloves with lights at the fingertips for easier viewing of objects in a box or at the back of a closet.	Devise a way to make a small figurine move across a tabletop as if by magic.

Selected References

Allen, Dorothea, *Elementary Science Activities for Every Month of the School Year.* West Nyack, NY: Parker Publishing Company, Inc., 1981.

————, *Science Demonstrations for the Elementary Classroom.* West Nyack, NY: Parker Publishing Company, Inc., 1988.

Appel, Gary, Margaret Cadous, and Roberta Jaffe, *The Growing Classroom.* Capitola, CA: 1985.

Banister, Dr. Keith and Dr. Andrew Campbell (eds.), *The Encyclopedia of Aquatic Life.* New York: Facts on File Publications, 1986.

Barhydt, Frances Bartlett, *Science Discovery Activities Kit.* West Nyack, NY: The Center for Applied Research in Education, 1989.

Blackwelder, Sheila Kyser, *Science for All Seasons: Science Experiences for Young Children.* Englewood Cliffs, NJ: Prentice-Hall, Inc., 1980.

Bruce, Katherine and Jessie Allen (eds.), *Science Experiments on File, Vol. 1.* New York: Facts on File Publications, 1989.

Cooper, Chris and Jane Insley, *The World of Science: How Does it Work?* New York: Facts on File Publications, 1986.

Cooper, Chris and Tony Osman, *The World of Science: How Everyday Things Work.* New York: Facts on File Publications, 1984.

Dempsey, Michael W. (ed.), *Illustrated Fact Book of Science, Vol. 2.* New York: Arco Publishing, Inc., 1983.

DiCanio, Dr. Margaret (ed.), *The Facts on File Scientific Yearbook.* New York: Facts on File Publications, 1988.

Downs, Gary and Jack Gerlovich, *Science Safety for Elementary Teachers.* Ames, IA: The Iowa State University Press, 1983.

Embry, Lynn, *Scientific Encounters of the Curious Kind.* Carthage, IL: Good Apple, Inc., 1984.

Evans, Peter, *Your World 2000, Planet Earth.* New York: Facts on File Publications, 1985.

Facts on File, *The History of Science and Technology.* New York: Facts on File Publications, 1988.

Hanks, Kurt and Jay Parry, *Wake Up Your Creative Genius*. Los Altos, CA: William Kaufmann, Inc., 1983.

Hutchins, Ross E., *Nature Invented It First*. New York: Dodd Mead and Company, 1980.

Lambert, David and Tony Osman, *Great Discoveries and Inventions*. New York: Facts on File Publications, 1985.

Levenson, Elaine, *Teaching Children About Science*. Englewood Cliffs, NJ: Prentice-Hall, Inc., 1985.

Mitchell, John, *The Curious Naturalist*. Massachusetts Audubon Society, 1980.

Nelson, L.W. and G.C. Lorbeer, *Science Activities for Elementary Children*. Dubuque, IA: Wm. C. Brown, 1980.

O'Hanlon, Maggie and Doreen Edmond, *Wild Animals of the World*. New York: Sterling Publishing Co., Inc., 1982.

Polley, Jane (ed.), *Stories Behind Everyday Things*. Reader's Digest Publications, Inc., 1980.

Ronan, Colin A., *Science: Its History and Development*. New York: Facts on File Publications, 1983.

Sawyer, Roger Williams and Robert A. Farmer, *Science Fair Projects*. New York: Arco Publishing, Inc., 1980.

Seabury, Debra L. and Susan L. Peeples, *Ready-to-Use Science Activities for the Elementary Grades*. West Nyack, NY: The Center for Applied Research in Education, 1987.

Sisson, Edith A., *Nature with Children of All Ages*. Englewood Cliffs, NJ: Prentice-Hall, Inc., 1982.

Sund, Robert B., Donald K. Adams, Jay Hackett, and Robert Moyer, *Accent on Science*. Columbus, OH: Charles Merrill Publishing Company, 1985.

Taylor, Ron, *The World of Science: Projects*. New York: Facts on File Publications, 1985.

Ward, Brian R. and Franklin Watts, *The Human Body*. New York: Franklin Watts Press, 1981.

Waxter, Julia B., *The Science Cookbook*. Belmont, CA: Pitman Learning, Inc., 1981.